SpringerBriefs in Applied Sciences and Technology

SpringerBriefs present concise summaries of cutting-edge research and practical applications across a wide spectrum of fields. Featuring compact volumes of 50 to 125 pages, the series covers a range of content from professional to academic.

Typical publications can be:

- A timely report of state-of-the art methods
- An introduction to or a manual for the application of mathematical or computer techniques
- A bridge between new research results, as published in journal articles
- A snapshot of a hot or emerging topic
- An in-depth case study
- A presentation of core concepts that students must understand in order to make independent contributions

SpringerBriefs are characterized by fast, global electronic dissemination, standard publishing contracts, standardized manuscript preparation and formatting guidelines, and expedited production schedules.

On the one hand, **SpringerBriefs in Applied Sciences and Technology** are devoted to the publication of fundamentals and applications within the different classical engineering disciplines as well as in interdisciplinary fields that recently emerged between these areas. On the other hand, as the boundary separating fundamental research and applied technology is more and more dissolving, this series is particularly open to trans-disciplinary topics between fundamental science and engineering.

Indexed by EI-Compendex, SCOPUS and Springerlink.

Ji Wu

Space Tourism

From Technology to Management

Springer

Science Press
Beijing

Ji Wu
National Space Science Center
Beijing, China

Translated by
Yongjian Xu
National Space Science Center
Beijing, China

ISSN 2191-530X ISSN 2191-5318 (electronic)
SpringerBriefs in Applied Sciences and Technology
ISBN 978-981-92-0882-1 ISBN 978-981-92-0883-8 (eBook)
https://doi.org/10.1007/978-981-92-0883-8

Jointly published with Science Press

The print edition is not for sale in China (Mainland). Customers from China (Mainland) please order the print book from: Science Press.

Jointly published with Science Press

This Springer imprint is published by the registered company Springer Nature Singapore Pte Ltd.
The registered company address is: 152 Beach Road, #21-01/04 Gateway East, Singapore 189721, Singapore

Foreword

Space tourism is a fact today, 68 years after the launch of Sputnik, thanks to private operators. It is still limited to a few tens of tourists per year, still limited to crossing the border of 100 km for most of them, and to short visits to the International Space Station for a few, but the number is already more than that of professional astronauts in one year and even private operators provide additional opportunities for government space agencies to fly their professional astronauts. The limits are designed by the very high cost, by the shortage of appropriate capacities and by the lack of proper regulations, but the doors are open, the dreams are at reach, based on the curiosity of all humans to discover the unknown. I can certainly measure that curiosity when I am walking in the streets of Paris, the number of tourists increasing by 10% every year.

I am among the dreamers looking for visiting space since long. I have tried so many times and failed so far. I was one of the first five French short-listed as candidate astronaut in 1977 during the first selection organized ever by the French Space Agency, CNES, but not good enough to go to space as a professional astronaut. I became DG of ESA as a back-up of not being an astronaut, accompanying the ESA astronauts to the bottom of the rocket taking them into orbit and offering each of them to exchange our positions, but none accepted to become DG. And since I left ESA in 2015, I am looking for a ride, but not rich enough to buy a ticket. Time is passing, but I continue to dream by reading the book written by Professor Wu Ji.

Space tourism is a fact today but Wu Ji draws in his book the dimensions of its future and the conditions of its development, describing the rationale and the paths for making space at reach for more and more humans. Not just a matter of equality among humans, but also a path toward a change in the behavior of individual humans for the benefit of the whole humanity. As a matter of fact, if space has changed the world in the last 68 years as of today, it has not changed the behavior of the humans which is quite opposite to it should be: humans are taking ownership of space after having taken ownership of planet Earth, while they are just very small parts of it. Humans have organized planet Earth in parts, while there is only one planet Earth for all of them. Humans consider their life limited in time and the planet Earth resources unlimited, while their objective should be to make the life of humanity unlimited on a finite planet Earth.

This is therefore the right time to take the opportunity of a space still without borders to open the eyes of as many humans as possible by offering them the possibility to visit space and to take some distance from the problems they have created on Planet Earth. As a matter of fact, the humans are at the same time the reason, the problem and the solution of all we are doing on Planet Earth and in space:

- the reason since the end objective is to make humanity sustainable,
- the problem since their behavior is opposite to the one it should be and
- the solution since the brain of humans is the only natural resource growing on Planet Earth.

Space tourism opens therefore a chance to change the mindset of humans before they organize space as they have organized Planet Earth. I can measure every day in Paris the benefits of tourism providing opportunities for more and more humans to meet peacefully but also the drawbacks of tourism not only in Paris but also in more and more places of Planet Earth. This is the reason why, space tourism must be carefully thought and properly organized and regulated from start in order to avoid overtourism followed by restrictions.

This is the objective of Professor Wu Ji, who is providing a well-thought approach of space tourism, based on his knowledge, experience and passion. Wu Ji is a great scientist, world class. I have learnt so much from him about science of the Universe. But he is much more than a scientist, he is a citizen of the world, sharing his passion all over the world. It is bold for a scientist to dare and address tourism and we should all thank him of doing it, and doing it as a rigorous scientist.

Wu Ji describes in this book all the mechanisms necessary for developing the capacities, for reducing the costs and for defining appropriate regulations able to multiply the chances for more and more humans to visit space, the only place with no borders so far. This fantastic objective will challenge the creativity of engineers, the interests of investors and the imagination of lawyers, with returns going much beyond tourism. The book provides the recipes, just do it and the only risk is an additional chance for humanity to survive.

4 October 2025

Former Director General of European Space Agency (ESA)
Paris, France
Jean-Jacques Dordain

Space Tourism: From Technology to Management

Space Tourism: From Technology to Management comprehensively explains the significance of space tourism to human beings' sustainable development, introduces the current technological capabilities and various feasible technological solutions, analyzes the potential market, and conducts a detailed cost analysis. In addition, it puts forward risk mitigation measures, discusses the design of space tourism services, and makes suggestions for the formulation of government policies. The book comprehensively covers the theoretical, technological, economic, and policy aspects of space tourism.

Book Features

- It is a comprehensive, unique, introductory book for space tourism touching upon various aspects, such as science, technology, economy, tourism service design, risk analysis, and policy.
- As a veteran space science leader, the author has abundant knowledge and experience in space science and technology, space transportation, and space mission planning, adding the authority in the text writing.
- This book studies the mentality changes after humans entered the space era, and discusses the growth point of space tourism in the broad background of shifting from traditional space endeavor to New Space. The book endorses that space tourism will bring about mentality changes.
- The author not only introduces the existing technologies but also provides technical solutions for future space tourism.
- The book is written in easy-to-understand language to explain complicated science and technical stuff to the general readers, which is inspirational not only to the readers in the academia but also to the general public as well.

This book is an excellent resource for researchers, academics, and graduate and undergraduate students in space science and technology who wish to understand space tourism. It also inspires private space tourism industry with its comprehensive account of all aspects relating to space tourism, including but not limited to technology solutions, tourism service design, etc. It is a public outreach source for the general public who are interested in space tourism.

Preface

The idea of writing a book on Space Tourism originated in 2019, a time to celebrate the 50th anniversary of the Moon landing by Apollo 11. At that time, one question lingered in my mind: 50 years have passed, why hadn't humans set foot on the Moon ever since?

After extensive research and reflection, I arrived at the following conclusion: the Apollo program was a political mission to gain an upper hand in the Space Race. Once this goal was accomplished, the program naturally lost its momentum and became unsustainable. To fulfill political objectives, the program was implemented with little regard for cost, leading to a vicious cycle of high expenses in government-led space programs. High investment entails high risks, which in turn demand extremely high reliability. The pursuit of high reliability further drives up costs. Thus, if failing to break free from this cycle of high costs, even with sustainable scientific objectives, manned lunar missions will not be sustainable, which explains why humans did not set foot on the Moon again in the past five decades.

However, about 10 years ago, a new paradigm emerged in space endeavor, which is distinct from traditional government-led efforts. This new approach (New Space), pioneered by SpaceX, emphasizes technological innovation and rapid iterative development. Unlike its predecessors, New Space is often driven by market demands. With such market-driven incentives, a self-sustaining and expanding industry independent of government mandates may take shape. This new industry is space tourism.

In the year 2020, when I was drafting the initial manuscript of this book, encouraging developments in space tourism began to emerge. Virgin Galactic, Blue Origin as well as SpaceX—the pioneer in reusable rocket technology—were accelerating in full speed their plans for space tourism. Approximately six months after the Chinese edition of this book was published in 2021, all three companies successfully completed their first commercial space tourism flights. In retrospect, we have every reason to view the year 2021 as the true inaugural year of space tourism.

As I am drafting this preface for the English edition, Blue Origin has conducted 12 commercial flights and Virgin Galactic has completed 7. SpaceX—through rapid iteration of its Starships—is poised to reduce the cost of launching payloads to Low

Earth Orbit (LEO) to approximately $1000 per kilogram. Undoubtedly, these progresses will significantly drive space tourism forward. It is worth noting that when the Chinese edition was written, cost estimates for LEO launches were somewhere around $3000–$4000 per kilogram. Given the rapid development of New Space in a matter of a few years, these figures no longer reflect reality. Nevertheless, I have decided to retain the original estimates in this English edition for two reasons: First, revising all cost-related analysis would alter the original discussions and conclusions; second, if readers find the initial discussions reasonable and accept the derived ticket prices, the significantly lower launch costs for the time being will further reinforce the validity of my projections. I am confident that space tourism tailored for the general middle class is on the horizon.

I would like to take this opportunity to reiterate the primary goal of this book: to show readers that space tourism is not elusive science fiction but a tangible reality. Moreover, space tourism is not merely joy-riding in space; its deeper significance lies in its potential to transform the space travelers and purify their minds. This will happen especially when viewing our home planet Earth from space. I believe that, like the vast majority of those who have traveled to space, the space tourists would return to the Earth as new persons—ones with a broader global perspective and a stronger commitment to sustainable development. They would cherish more their family and friends, their own city and country, and the entire planet at large. This transformation is akin to the profound personal change experienced after climbing Mount Everest and returning to base camp, though the impact of space travel is likely to be more intense and enduring.

Technological advancements today are rapid. Most of the technologies discussed in this book have a high technology readiness level. I want to draw your attention especially to the in-situ utilization of lunar resources, which requires further development. However, this is not a theoretical challenge but rather a matter of technical and engineering implementation. Therefore, in the final chapter of this book, I urge governments of space-faring nations and venture capitalists to prioritize investment in key technologies that may otherwise hinder future progress. With dedicated and concerted effort and sufficient funding, a new era in which thousands of ordinary people travel to space for tourism will arrive sooner than we can imagine.

Beijing, China
September 2025

Ji Wu

Contents

About the Author

Ji Wu is the President of the Chinese Society of Space Research (CSSR), former Director General of the National Space Science Center (NSSC) of the Chinese Academy of Sciences (CAS), Full Member of the International Astronautics Academy (IAA), IEEE Fellow and Foreign Member of the Russian Academy of Sciences. He once served as the Vice President of the Committee on Space Research (COSPAR) (2010–2018), and he is an awardee of the International Cooperation Award of COSPAR in 2022. He was also deeply involved in several international space science missions, such as the Double Star Program with the European Space Agency and the Yinghuo-1 Mars mission with Russia. He was in charge of the Strategic Priority Program on Space Science (Phase I), which includes the following scientific missions: the Dark Matter Particle Explorer (DAMPE, or Wukong), the Quantum Experiments at Space Scale (QUESS, or Micius), the Hard X-ray Modulation Telescope (HXMT, or Insight), and the Shijian-10 recoverable satellite mission.

Chapter 1
Introduction

Before the formal discussion, we must make clear a critical fact: The greatest challenge of space tourism lies in accelerating a spacecraft to the first cosmic velocity (approximately 7.9 km/s) to overcome the Earth's gravitational pull. Thus, the feasibility of commercial space tourism hinges on humanity's ability to achieve this velocity in a cost-effective and sustainable manner. Only after overcoming this challenge can we enable routine travel between the Earth and space, making commercial space tourism truly feasible. Therefore, space tourism is not merely a technical challenge. It is a techno-economic issue.

Venturing into space is not as easy as traveling on the Earth, where our movement is literally unrestricted. For prehistoric humans, travel means walking on foot. With the rise of civilization, travel increasingly relies on carts and boats. Even today, modern air flight merely speeds up movement—it does not free us from the Earth's gravity. We remain confined to two-dimensional (2D) movement, constrained by the planet's gravitational pull. Space travel, however, introduces a third dimension: escaping from the Earth's surface to fly upward into space. Adding this extra dimension means that we have to overcome the Earth's gravity. Newtonian mechanics dictates that an object in circular motion generates centrifugal force proportional to the square of its velocity. Once this outward force balances the Earth's gravitational pull, orbital flight becomes possible. This equilibrium defines the first cosmic velocity as mentioned above. Should the centrifugal force exceed the Earth's gravity, the spacecraft will ascend to higher orbits, eventually reaching the second cosmic velocity (approximately 11.2 km/s) to completely leave the Earth behind.

When examining our solar system, we face an undeniable fact: The Planet Earth is the most difficult solid body to escape from [1]. If the Earth's gravitational acceleration is defined as 1g, Venus comes closest at 0.905g, followed by Mars (0.379g) and Mercury (0.378g)—all weaker than Earth's. The outer planets are gas giants with no solid surface, and even the moons of all planets in the solar system have lower gravitational acceleration. Thus, the Planet Earth stands alone as the most challenging solid body in the solar system in terms of venturing into space. This reality underscores that space tourism is not an easy "get-away." The primary

J. Wu, *Space Tourism*, SpringerBriefs in Applied Sciences and Technology,
https://doi.org/10.1007/978-981-92-0883-8_1

techno-economic obstacle is achieving the first cosmic velocity in an efficient and affordable manner to overcome the Earth's gravity. Otherwise, space tourism still remains at the science fiction level.

Some might argue that humans have already entered space. Haven't they? The US astronauts landed on the Moon 50 years ago, and a few tourists have visited the International Space Station (ISS). Therefore, it is apparent that the technical issues have been resolved. This argument has solid grounding. Technically, the first cosmic velocity has been reached by Sputnik, the first artificial satellite, making it not a mere equation on the paper. But if we regard government programs or individual billionaires' ISS trips as viable space tourism, we would be completely wrong. Government-funded programs, such as the Apollo program, are politically driven, allowing unlimited budgets until success. Yet such programs are inherently unsustainable. Once the goals are met, the government will lose motivation and suspend the program, which explains why there are no manned lunar missions after the Apollo program. Similarly, a handful of billionaires cannot sustain the space tourism economy.

Sustainable space tourism requires a market where participation from ordinary people grows steadily, even if the participation is initially limited to middle and high-income groups. Only then can a self-reinforcing space tourism economy emerge. To achieve this, we need to further improve the technology solutions developed by the government to overcome the Earth's gravity, a fundamental techno-economic issue as mentioned above. We have added sustainable and self-reinforcing development as the premise for the materialization of space tourism.

If we disregard economic feasibility, space tourism has come into shape for two decades since 2001, when American businessman Dennis Tito became the first paying visitor to ISS [2]. Yet progress has stagnated. We haven't seen any substantial advancement ever since. In 2021, National Aeronautics and Space Administration (NASA) raised the price for an ISS visit to $57 million per passenger. This shows that a handful of ultra-wealthy customers cannot sustain the space tourism industry. True space tourism must be discussed within the framework of sustainability, i.e., making it accessible for ordinary people. "Ordinary" refers tentatively to middle and high-income groups, instead of billionaires or average wage-earners. A practical benchmark would be quantitative: only when thousands or even tens of thousands of tourists enter space annually can we declare that the era of space tourism has truly arrived.

Another critical question must be addressed: What is the actual purpose and content of space tourism? So far, the public perception of space tourism remains fixated on floating in microgravity. They tend to regard the state of weightlessness as the sole attraction for tourists. This is a misconception. In this book, we will explore how humans' perception and mentality may change upon entering space, as our worldview is inherently shaped by the environment we live in. When frequent, affordable Earth-space travel becomes a reality, this change of environment will inevitably reshape humans' perception and mentality. Moreover, space tourism won't be confined to space stations or space hotels in the Low Earth Orbit (LEO). It will extend to spaceship travels to the Moon, the Moon's surface, and eventually to other planets or moons in the solar system. Thus, microgravity is not the default setting for space tourism. It should be noted that once in situ resource utilization (ISRU) technologies become mature, making fuel extraction, production, and refueling in space possible, the interplanetary voyages may experience continuous acceleration. It will not only shorten travel time but also generate artificial gravity, providing more convenience for tourists. In a word, floating in a weightless microgravity environment is hardly the standard setting for space tourism. In the future, the microgravity environment may be merely a short transitional phase rather than the norm for space tourists.

Guided by these fresh perspectives, this book will begin with humanity's journey into space, and examine, through a techno-economic lens, why space technologies have long been confined to government programs rather than commercial space tourism. We will then explore the ultimate purpose and impact of space tourism from the evolutionary perspective of mentality changes brought by space tourism, so as to increase public awareness of its significance. Subsequent chapters will analyze related critical infrastructures, assess the economic feasibility of various technologies, and evaluate market dynamics and risks associated with space tourism. The concluding chapters will focus on the service design and immersive experience of the tourists, and offer policy recommendations for governments.

This book is written for general readers. Those who find technical sections overly detailed may skip over the details and get familiar with the chapter conclusions. For readers who have the fundamental knowledge of these technologies, they are encouraged to study the referenced materials for extended reading.

References

1. A.E. Roy, D. Clarke, *Astronomy: Principles and Practice*, 4th edn. (CRC Press, Boca Raton, 2003)
2. H. Feldman, *Dennis Tito: The First Space Tourist* (PowerKids Press, New York, 2002)

Chapter 2
The Space Age

Breaking free from the Earth's gravitational pull to enter Low Earth Orbit (LEO) marks the beginning of the Space Age. However, more expeditions are waiting for us to explore the far reaches of space, such as the Moon, Mars, and beyond. For the time being, we are at the most elementary stage of the Space Age.

2.1 The Beginning of the Space Age

The Soviet scientist Konstantin Tsiolkovsky was the first to create the initial theory on how to enter space. He originally calculated the first cosmic velocity to achieve orbital flight around the Earth, and the second cosmic velocity to leave the Earth and enter the solar system. In addition, he proposed a multi-stage rocket design. It was the German engineer Wernher von Braun, chief designer of Nazi Germany's V2 missiles during World War II, who actually developed a vehicle capable of spaceflight. On the basis of the V2 missile engine, the Soviet Union developed a launch vehicle capable of launching an artificial satellite. In October 1957, the Soviet Union successfully launched the first artificial satellite, Sputnik-1, ushering the Space Age. In April 1961, the Soviet Union launched the first manned spaceship, sending the first cosmonaut, Yuri Alekseyevich Gagarin, into the Earth's orbit. Despite orbiting the Earth just once, the spaceship heralded the dawn of human spaceflight.

From the late 1950s to the late 1980s, against the backdrop of the Cold War, space activities were in a state of intense competition [1]. The Soviet Union and the United States, as rival superpowers, invested enormous human, material, and financial resources in spaceflight to excel each other. In this process, they achieved remarkable feats that seemed almost impossible given the technological limitations of the time. These included the first successful human spaceflight, accomplished within 4 years after the launch of the first artificial satellite, and the Moon landing, achieved in less than 12 years after the launch of the first man-made satellite.

J. Wu, *Space Tourism*, SpringerBriefs in Applied Sciences and Technology,
https://doi.org/10.1007/978-981-92-0883-8_2

Additionally, numerous probes and spacecraft were launched to explore the Moon, Venus, and Mars. These space activities have significantly advanced various fields of space science, such as space physics, space astronomy, Earth observation, planetary science, and microgravity and space life sciences. The technologies developed in the space race have also given rise to many new applications, such as global satellite communications, a large number of remote sensing applications on military and civilian satellites, and the construction of global navigation satellite systems.

While the Space Age originated from the rivalry between two superpowers during the Cold War, it ultimately ushered humanity into the Space Age. Then, how should we define the Space Age?

First, in the Space Age, mankind has the technological capability to overcome the Earth's gravity and accelerate a spacecraft to the first cosmic velocity to enter orbital space. This capability is as pivotal as James Watt's improvements to the steam engine, which enabled large-scale machine production and transitioned humanity from an agricultural to an industrial society. Although the cost of overcoming the Earth's gravity to go into space is still very high, significant breakthroughs in theory and technology have been achieved in the meantime. The primary challenge now is to reduce costs and enable more frequent trips between the Earth and space.

Second, in the Space Age, space technology has become an integral part of our daily lives rather than only in science fiction. National security, communication and navigation, weather forecasting and even our daily lives are all directly or indirectly influenced by space science and technology.

Last but not least, in the Space Age, our mentality and perception began to change [2]. Before entering space, the scope of human activities was confined to the surface of the Earth. Through scientific observation and study, human beings were able to realize and understand that the Earth is round in shape, that the Earth is one of the eight planets in the solar system, and that the Moon is a natural satellite of the Earth and other knowledge in astronomy. However, all these findings stemmed from theoretical studies or mathematical deductions. Even an airplane's flight altitude, which is less than 20 km, is minimal compared to the Earth's radius of 6371 km. Thus, humans are essentially confined to the Earth's surface.

However, human beings have entered space and have been able to fly continuously in the Earth's orbit. In addition, human beings can travel from the Earth to the Moon and look back at the Earth from such a long distance. All these achievements began to change human beings' mentality and conceptions, and deeper considerations were given to the sustainable development of ourselves, as well as to the relationship of the Earth's environment and civilization. Especially after 2006, through the mirror of astronomical telescopes in orbit, human beings have discovered thousands of planets outside the solar system. It has dawned upon us that human beings might not be the only intelligent life in the universe, which has led to discussions about how to communicate with extraterrestrial intelligent life.

A number of conceptual changes have been triggered by mankind's adventure into space, including the rise of the environmental movement in the 1960s and the sustainable development debate in the 1970s. However, these very important messages from space were comprehended only by very few astronauts, especially the 24 astronauts of the Apollo Program. It was they who bypass us these messages [3]. But mankind has not paid enough attention to these messages, and some messages have even been ignored.

In the near future, we anticipate that more people will be able to go into space, especially ordinary people who will go into space as space tourists and directly receive the messages from space. This will be one of the primary purposes of space tourism. By that time, it can be said that mankind has entered the Space Age in a real sense, and the era we are currently living in can be deemed as the initial stage or the beginning of the Space Age.

2.2 Characteristics of the Space Age

Since the technologies enabling mankind's access to space originated from national defense, the space industry from the outset has been closely connected with national defense. In a sense, the space industry implies secrecy and military applications. Besides the two superpowers, i.e., the United States and the Soviet Union, even the space industries in other countries with launch capabilities also have the same connection. Therefore, in the Space Age, it was inevitable that the space sector and the government were tightly tied together, which is mainly reflected in the following aspects.

First, the vast majority of space missions are government programs, such as defense missions, missions for public service, and scientific exploration missions. In the early days of the Space Age, even the communication and broadcasting satellites were all initially funded by the government for development and testing. They were then transferred for commercial use when practical applications became feasible.

Second, the main space players in each country are large state enterprises, or even governmental institutions and research institutes. The main source of funding for these players comes from government-dominated missions.

Third, competition is limited due to the fact that missions are planned and organized by the government. This often results in a balance between two major state companies, leading to a monopoly on pricing and high mission costs.

Due to these very inherent features, the mission developers have coincidentally placed a high priority on ensuring mission success, regardless of the cost. The main reason is that funding comes from the government, i.e., the taxpayers, and mission failure was almost a "crime" against the people. Over time, this tradition of zero tolerance to failure became the accepted standard for space missions. With the rise of commercial space, practitioners from government agencies and the space sector carried this tradition over to the new commercial space industry. However, commercial space is funded by users who will make new requests based on their needs, particularly within the constraints of cost. Space tourism is the typical commercial space market. Space tourism's tolerance for risk and associated risk mitigation measures will be discussed later in Chap. 10.

The lack of competition mechanism and the zero tolerance to failure lead the development of government missions into a vicious cycle, i.e., mission design with high reliability—a large number of ground tests—long development cycle—high cost, adding pressure for mission success—mission design with even higher reliability. Therefore, without the commercial demand for cost reduction and the regulation by the invisible hand of the market economy, it is impossible to make ordinary people's space tourism a reality. This shows once again that space tourism is not only a technical problem, but also a typical complex techno-economic problem.

2.3 The Beginning of Space Tourism

While space activities in various countries remain predominantly government-funded, the International Space Station (ISS) has indeed hosted seven self-funded visitors since 2001. These visitors may be considered as the pioneers of space tourism. Their profiles are as follows:

Dennis Tito (American businessman): He is the first private visitor to the ISS. He prefers the terms "private space explorer" or "private astronaut" rather than "tourist," emphasizing his equivalent status with official astronauts. After training at Russia's Gagarin Cosmonaut Training Center, including microgravity aircraft training, centrifuge practice, and supersonic jet exercises, he arrived at the ISS aboard Soyuz TM-32 on April 28, 2001, and stayed there for 9 days.

Mark Shuttleworth (South African businessman): He is the second private visitor to the ISS. After 8 months of specialized training and rigorous medical evaluations at the Gagarin Center, he flew on Soyuz TM-34 on April 25, 2002, for an 11-day stay. He framed his mission as an educational inspiration for South African youth.

Gregory Olsen (American businessman): After completing 900 hours' training at Gagarin Center, he reach ed. the ISS via Soyuz TMA-7 on October 1, 2005, and stayed on the ISS for 11 days. He participated in ESA-led space medicine experiments and conducted amateur radio sessions with students in New Jersey and New York.

Anousheh Ansari (Iranian-American businesswoman): She is the co-sponsor of the Ansari X Prize. As of January 2021, she is the only female self-funded space visitor. She was launched into space by Soyuz TMA-9 mission on September 18, 2006 and stayed at the ISS for 12 days. She took part in four life science experiments led by ESA, including studies on mitigating astronauts' back pain.

Charles Simonyi (Hungarian-American): He is the former Chief Architect and software engineer of Microsoft. He is the only two-time private visitor to the ISS (Soyuz TMA-10 on April 7, 2007 for 15 days; Soyuz TMA-14 on May 26, 2009 for 14 days), holding the record for longest cumulative stay by private astronauts.

Richard Garriott (British and American businessman): He is son of NASA astronaut Owen Garriott. He visited the ISS by Soyuz TMA-13 mission launched on

October 12, 2008, for a 12-day stay. He tried to promote commercial space through his efforts.

Guy Laliberté (Canadian businessman): He is the founder of Cirque du Soleil. He is the last private visitor to the ISS to date. He was launched into space by Soyuz TMA-14 on September 30, 2009, for a 12-day journey. He coined his journey "Poetic Social Mission" with the theme on the "Moving Stars and Earth for Water". The centerpiece event took place on October 9 from space and in 14 cities simultaneously around the world, beginning at 8:00 p.m. (ET). Laliberté hosted the 120-min event at the ISS, and artists and celebrities worldwide were able to look up at the stars and took part in a variety of activities on Earth. These celebrities expressed their commitment to water conservation and spoke highly of this vital natural resource.

All visitors were organized by Virginia-based Space Adventures, Ltd. The initial pricing was $20 million per person, including a 6-month training and a ~10-day stay at the ISS. By 2020, NASA's updated rate reached $55 million, with spacewalks costing an additional $15 million and requiring 8 extra training days.

It is clear that such costs are not affordable for ordinary tourists. As these ISS visitors emphasized, they are private astronauts rather than tourists. Notably, all the space facilities these private visitors used for entering space, staying on the ISS, and subsequent reentry are government-owned space assets. Funding to construct and maintain these facilities comes from national space programs rather than the commercial space sector. Therefore, their visit to the ISS is not deemed a commercial act. The market mechanism for commercial space doesn't have a role to play in these cases, which explains the persistently high costs.

The true era of space tourism has not yet arrived.

References

1. D. Cadbury, *Space Race: The Epic Battle Between America and the Soviet Union for Dominion of Space* (HarperCollins, New York, 2007)
2. J. Wu, What change may take place after we entered space era. Bull. Chin. Acad. Sci. **36**(10), 14 (2021)
3. R. Poole, *Earthrise: How Man First Saw the Earth* (Yale University Press, New Haven, 2008)

Chapter 3
Enlightenments from Space

Before entering the Space Age, humanity had already gained a preliminary understanding of itself and the universe beyond. The scientific community had given us a nearly accurate picture of the Earth and the solar system, particularly after Isaac Newton formulated his Three Laws of Motion. However, some philosophical questions are still haunting human beings, such as Who are we? Where do we come from? Where are we going? In ancient times, Chinese Daoism also attempted to answer these questions through the philosophical concept of "Harmony between Humanity and Nature" (Tiānrén Héyī).[1]

But it wasn't until modern times, when we finally ventured into space, that these questions took on new weight and truly began to resonate deeply within us. These reflections are due to the changes in our living and working environments, and show the new message from the new environments to humanity. Just as primates descended from trees and began walking upright, or as humans first sailed from land into the ocean, the new environment has triggered profound changes in our mentality and societal structures. This chapter will explore and discuss how these changes occurred and what enlightenments they have brought to humankind.

The following sections will introduce the three enlightenments from space in chronological order and present the significant events associated with these enlightenments.

[1] An integral facet of the Chinese worldview, it shines light on people's basic understanding of the cosmos and their relationship with it. The idea stresses global unity, that human beings, alongside other living creatures, coexist with nature and constitute the cosmos, the biggest shared community of life. At the same time, it also recognizes the uniqueness of human beings for their ability to understand and exploit the laws of the universe.

J. Wu, *Space Tourism*, SpringerBriefs in Applied Sciences and Technology,
https://doi.org/10.1007/978-981-92-0883-8_3

3.1 The First Enlightenment

The first enlightenment occurred when human beings first ventured into space, and it was brought by the Soviet cosmonaut Yuri Gagarin, who was the first human to orbit Earth. His historic flight lasted just more than 1 h, and Gagarin's schedule was tightly planned with prescribed procedures and tasks, leaving him almost no time to gaze through the capsule window to enjoy the breathtaking view of our Planet Earth. Even then, he exclaimed from space, "It's beautiful. What beauty!"[2] The astronauts and taikonauts who followed, upon finishing demanding work, had leisure time to enjoy the beautiful Earth. For instance, when interviewed by reporters after returning to the Earth, Chinese Taikonaut Jing Haipeng said, "I slowly edged to the window to have a look of the Earth. Language fails to reveal the beauty of the brownish yellow land, the majestic mountains and the ribbon-like rivers. At that moment, I couldn't help thinking about the boundlessness of the universe, the insignificance of individuals, and the greatness of our nation, and I was so proud of being a Chinese."[3] When delivering a lecture to students, Ms. Liu Yang, the first Chinese female taikonaut, said, "it is true that the Earth is round in shape". In order to better appreciate the scenery of the Earth, a special 360-degree viewing platform has even been built on the ISS. In December 2018, after performing a spacewalk, Anne McClain, a female American astronaut, exclaimed:

> I feel like I was very connected to Earth, like it was mine and there was some ownership there, not only my hometown or my city or what I was familiar with, but the entire Earth. I felt an ownership and a kindred spirit with every part of it. You do feel this closeness to every human on Earth. You have this really distinct recognition that every single person you will ever meet, you have more in common with than you do different, no matter what.[4]

When an astronaut is launched beyond the atmosphere, he or she enters the realm of space upon crossing the 100-km threshold known as the Kármán line. He will see the boundless pitch-dark universe studded with galaxies and stars, and the Earth's horizon, with distinct curvature, is embraced by Earth's fragile thin atmosphere. Beneath his feet, he will also see the fast-changing cloud patterns, blue oceans, yellow and green continents, as well as white polar ice caps. This vantage point evokes profound thoughts about the greatness and smallness of humanity. Our greatness manifests in our scientific and technological achievements that enable us, intelligent creatures, to behold our home planet from such a high altitude in space. Yet in comparison with the mother Earth and the great Nature, our smallness becomes equally evident there, since nearly all traces of human civilization disappear, reducing us to just one of the countless species in the Earth's biosphere. The astronaut comes to realize the unity of human beings: our common grounding far outweighs differences, binding us as one species.

---

[2] *BBC Sky at Night* [2021-03-25]. https://www.skyatnightmagazine.com/space-missions/yuri-gagarin-first-human-in-space.

[3] *Securities Times*. Chinese astronauts in space classic footage playback: Jing Haipeng space photography. [2021-01-20]. https://finance.qq.com/a/20120617/000892.htm.

[4] NASA Astronaut Anne McClain Tells of Her First Spacewalk with Earth Awareness. [2021-01-20]. https://k.sina.com.cn/article_2206258462_m8380d51e03300l5uj.html?from=science.

Orbital flight awakens deeper existential awareness and compassion. Upon seeing the thin atmosphere, the astronaut can feel its vulnerability. It seems that any perturbation from space might strip the fragile atmosphere away, exposing all terrestrial life to danger or even annihilation. This vision triggers philosophical thinking about earthly existence: the loved ones, the friends, the communities, the cities, and home country, which all rely on this precarious biosphere. Clouds, oceans, glaciers, and even deserts reveal sublime beauty, making us recognize ourselves as the Earth's children who must protect this cradle from destruction. Destroying the planet Earth constitutes an intergenerational crime, a self-destruction for mankind. Harboring these kinds of sentiments, the returning astronauts consequently become environmental advocates, champions of international cooperation, and ardent supporters of world peace and sustainable development.

Between 1961 and 1968, these intensified messages from space gradually captured public attention. Discussions emerged about obtaining the full picture of the Earth from further distances, and the public began to imagine how the Earth would look like there. There were two technological limitations at that time. First, sending satellites into high orbit required unprecedented launch capabilities. Second, digital photography was still elusive then. Only low-resolution images could be downlinked to the ground, or the photos could only be brought back to Earth by the reentry capsules of recoverable satellites. Things did not change much until 1968, when Apollo 8 flew beyond Earth orbit and circled the Moon for the first time.

3.2 The Second Enlightenment

The second enlightenment occurred in late 1968, when the Apollo 8 astronauts became the first humans to reach the lunar orbit. Originally, the mission had no plans to photograph Earth. All the training before launch and mission planning focused solely on the observation of the Moon. When Bill Anders happened to see the Earth rising above the Moon's horizon through the capsule window and wanted to take a photo, Commander Frank Borman's first reaction was: "hey, don't take that, it is not scheduled." Of course, this doesn't stop them as the first humans to receive the revelations from space from such a long distance away from the Earth. Bill Anders becomes the man behind the photo *Earthrise* (Fig. 3.1) which became extremely popular around the world [1].

When you venture slightly farther from the Earth to Geostationary Earth Orbit (GEO) to look back, or view the Earth from the Earth-Moon Lagrange L1 point 320,000 km away, or from the lunar orbit/surface at 380,000 km, you will realize that mother Earth is a planet, a solitary sphere spinning in the immense darkness of the universe. The areas bathed in the Sun are extremely beautiful, and we could see a bluish-white light; other areas shrouded in the darkness of night are illuminated by

Fig. 3.1 *Earthrise* taken by Apollo 8 astronaut Bill Anders on December 25, 1968

thunderstorm lightning and city lights. During solar storms, you might witness auroras dancing over the poles. From there, you can't tell the national borders and you will not hear any arguments. What remains visible are only oceans, continents, polar ice, and drifting clouds. When there are few clouds, the Earth reveals itself as an extremely beautiful blue planet.

This perspective inevitably makes you realize that the Earth we are living on is a planet in the solar system just like Mars or Venus. The only difference is that the Earth happens to be in the habitable zone with the right distance to the Sun. With the right temperature, the liquid water persists, which is the reason why the Earth takes on the blue and white color. This delicate equilibrium is both accidental and inevitable. It is accidental in that the Earth's temperature is conducive to the formation of water, which allows the emergence of life. It is inevitable in that after billions of years of evolution, intelligent beings will certainly arise. To the best of our knowledge, humanity represents the only intelligent life in the solar system. On this planetary oasis, superficial distinctions like skin color pale against our shared identity as human beings. We have learned to use tools and language, and we have created colorful cultures and advanced science and technologies. Therefore, we bear the sacred responsibility to nurture rather than obliterate our civilization's legacy. It is an unprecedented major responsibility for us. These collective epiphanies constitute the second enlightenment from space. In a word, we are responsible for the sustained development of our civilization.

The impact of this enlightenment dwarfed the first, making people begin to understand the significance of the various messages provided by the first enlightenment, and fundamentally reshaping humans' self-perception. Like ants that move in two-dimensional directions, human beings suddenly comprehended the significance of the vertical dimension and gained cosmic consciousness as we transcended terrestrial boundaries.

From 1966, there was an increasing public demand for a whole-Earth picture. The famous *Earthrise* photo taken by Apollo 8 in December 1968 started to fundamentally change humanity's perception of the Earth and ourselves, triggering discussions in academic writing, literary works, and movies. The discourse centered on existential questions: as the sole intelligent creature on the Earth, what should we do to ensure our sustainable development on the Planet Earth? When the Club of Rome published *The Limits to Growth* in 1972, articulating the constraints of Earth's carrying capacity, these discussions reached the zenith. It is reasonable to say that these discussions and associated mentality changes are the direct products of the Space Age, a milestone for the progress of civilization [2, 3].

The second enlightenment will be more profound if we can view the Earth from the lunar surface on a lunar night. All six Apollo mission landings occurred during lunar morning or afternoon when half or less than half of the Earth's disk is illuminated by the Sun. There are reasons for this arrangement. First, it is due to the energy constraint. The Sun provides the necessary energy supply. Second, it is due to the temperature constraint. At the lunar noon, the temperature exceeds 130 °C. If we can solve the energy supply issue and choose to land on lunar nights, we will see that more than half of the Earth's disk is illuminated by the Sun, and during the two Earth days in the middle of a lunar night, we will see a fully sunlit Earth dominating the lunar sky. It should be an extraordinary view. Although photographs of the Earth fully illuminated by the Sun can be taken by satellites in synchronous orbit or by probes flying into deep space, there is a big difference between that and seeing the view with naked eyes. In that scenario, I believe that space would undoubtedly convey more information to us.

3.3 The Third Enlightenment

The third enlightenment is in the wake of the second. It is built directly upon the newly acquired frontier knowledge of the universe through the progress of space science and technology, particularly through robotic space missions that revealed unprecedented details about our solar system and the universe beyond.

As humans came to regard themselves as the only intelligent species among countless others in the solar system, they began to discuss the possibility of extraterrestrial intelligence. Early efforts focused on our solar system, sending probes to the Moon, Venus, and Mars. But when the Viking mission landed on Mars in 1976, only to find a barren and lifeless surface, our hopes of discovering intelligent neighbors within our solar system vanished. Though discussions are still going on, they are largely confined to science fiction. In 1995, astronomers detected the first exoplanet using Doppler velocimetry measurements. It provides scientific observation grounding for those who believe the existence of extraterrestrial intelligence.

Thanks to telescopes both on Earth and in space, including NASA's Kepler space observatory launched in 2006, we have identified over 4000 exoplanets, and many of them are within the habitable zones. These discoveries suggest that nearly every star in the universe hosts orbiting planets. Therefore, for trillions of stars, it seems statistically inevitable that some Earth-like planets must have nurtured intelligent life. Thus, humanity is no longer the only intelligent life, and we are almost certainly not alone. This is the third enlightenment from space. The existence of extraterrestrial intelligence makes the question of whether to contact the aliens urgent.

This enlightenment deeply influenced the worldview of *The Three-Body Problem* trilogy written by Chinese sci-fi writer Liu Cixin [4]. In his narrative, different civilizations compete ruthlessly for resources, such as energy, and they even resort to annihilation through "dimensionality reduction attack." This kind of thinking may be very conducive to science fiction storytelling, as it creates suspense, tension, and drama, and it is also a way of translating the current level of development of human civilization into a possible extraterrestrial civilization. But this may not be the case. First of all, regarding energy, no civilization will ever consume it until it is depleted. What's more, energy is ubiquitous in the universe, and there is no need to travel to another star system with intelligent life to plunder energy resources and destroy the civilization there. Secondly, in the case of highly civilized beings, they must have acquired the conception of sustainable development of civilization in the process of setting off from their home planet. Like human beings who have received the second enlightenment from space, the moment they step out of their planet, they will first realize that protecting their homeland is the most important thing, and then they will be gradually inspired to find new civilizations and learn new knowledge. Therefore, the purpose of communication between different civilizations should mainly be mutual understanding and cooperation. With this in mind, they will discover and accommodate the differences of civilizations that have evolved under different environments, to learn the good things that don't exist in one's own civilization, and to constantly enrich one's own science and technology and cultural knowledge, in a bid to achieve the purpose of improving one's own development. Space colonization only happens when mankind takes the initiative to discover planets that are very suitable for human survival or even more habitable than the Earth's environment, and is willing to enter into that biosphere. But this is far beyond the scope of the discussion of the third enlightenment from space.

3.4 Elusiveness of Collective Human Memory

Humanity collectively remembers disasters they have experienced in person, such as plagues and wars. But these memories and perceptions will fade away unless the conditions that triggered the evolution persist. A good example is that the Moon landing is fading away from public memory.

In 1999, a US government survey, prior to its new round of lunar mission planning, revealed that 10% of respondents believed the Apollo program was a hoax. Three decades had bleached away the living experience, turning history into conspiracy theory. The zenith of planetary awareness and sustainability debates in the 1970s and 1980s triggered by viewing the Earth from afar has faded, as most developed countries in the world reverted to fighting for their own narrow interests. Even the USA, once the vanguard of cosmic thinking and campaign of advanced technology, succumbed to extreme nationalism and exceptionalism. This change precisely reflects the absence of the collective memory of humankind.

This saddening phenomenon has roots in hard reality: after more than 50 years, mankind has never set foot on the Moon since 1972. The conditions for the Apollo Program and its revelations no longer persist. The Apollo astronauts in their 1980s are still with us, and they are working hard to share their stories to appeal to the public, only to find diminishing social impact. Their voices are drowned in the engine sound of chips, quantum technology, and AI. Our attention is gradually turning inward rather than outward. We are focusing on microscopic simulation and creating virtual satisfaction at the micro level, rather than on the macro level, such as setting sail for the solar system and space beyond. I want to particularly mention the efforts of Elon Musk, the American businessman, who is still trying to bring humanity's attention back to the macro level.

However, in terms of the evolutionary trajectory of an intelligent species, it is inevitable to move from two-dimensional to three-dimensional. The apes descended from trees to walk upright and then evolved into humans. After that, they gradually migrated to all continents. They ventured from the continents into the vast oceans, and began the great voyage, an era of great discovery. In this process, they built the global market. In recent history, they invented the airplane to conquer the sky, so that human beings moved unprecedentedly faster, creating a new world pattern. The impact of these changes on humanity is enormous. Up to now, however, humankind has basically remained two-dimensional, confined to the Earth's surface. In a broad sense, even aerospace flights and orbital space stations are two-dimensional in that they have not really ventured out into the solar system. The Apollo program, as many people have experienced, may be regarded as only a fleeting strong pulse before a major evolutionary change in the long history of human evolution. Substantial changes would take place when tens of thousands, hundreds of thousands, or even millions of people would continuously travel into space and return to the Earth, visiting the Moon, or even more distant bodies, such as Mars. What can we call this scenario? Obviously, it is space tourism.

3.5 Significance of Space Tourism for Human Evolution

The evolution of mankind is characterized by adaptation to changes in the environment and an increase in knowledge. The descent from trees marks the farewell to the apes. Moving from the Earth's surface to space would be one of the most remarkable migrations for the human species, which will deliver a profound impact on human evolution.

However, we have yet to find a place in our solar system that is more favorable than the Earth. In addition, the current technology is not advanced enough to transform a barren celestial body into a place that is more suitable for dwelling than the Earth. What's more, immigrating to a celestial body that is completely incomparable to the Earth's environment and then confining ourselves once again to a single planet is certainly not the purpose of our venturing from the Earth's surface into three-dimensional space. Therefore, migration to Mars does not represent the technology direction of our evolution, and is not the future goal of our evolution either. Space tourism, on the other hand, is different. In terms of technology, the relevant technologies have basically been materialized. In terms of the direction of evolution, it represents the trend of human beings to leave the surface of the Earth and venture out into the solar system. As for the destinations, this will be a major issue for future space tourism. When lunar tourism becomes frequent enough, maybe we can visit other planets and their moons in the solar system. For example, we can visit Titan to have a close-up view of the beautiful rings of Saturn, or we can build an orbital space city for tourists at the Earth-moon Lagrange point, allowing humans to continuously travel back and forth between the space city and the Earth. In any case, mankind will not be bound to one planet again, and will be constantly traveling between the planets in our solar system. Thus, even if there were so-called settlements on celestial bodies beyond the Earth, the scenario is less likely to happen where a generation or even generations will never leave their settlements even if there are many people who would be born there.

Therefore, space tourism is deemed as a necessary stage in mankind's evolution, which will certainly bring about mentality changes in mankind. Accordingly, we cannot regard space tourism as only a techno-economic problem, it is an issue for the future development of mankind. Because these mentality changes will better serve the human beings, making them realize that their home planet is the only planet in the solar system that can support living even with its pristine environment, and that the Planet Earth is so beautiful. When you leave the Earth and travel through the solar system, you will definitely call it home, and you will wish it all the best, free from fighting and environmental destruction. This kind of change in perception and mentality will be the most attractive and far-reaching part of space tourism, which can even be regarded as the beginning of the future evolution of mankind. Just like when you have been to the Polar region or to China's Tibetan plateau, your soul will be purified, it is the same when you have been to space and looked back at your home planet there. For all of us whose movement are still confined to the surface of the Earth, we can imagine what we will experience spiritually.

3.6 Manned Space Exploration Vs. Space Tourism

Manned lunar exploration was suspended for more than 50 years after the Apollo program. But recently, we can see signs of recovery. The US government officially launched the Artemis Program in 2019 to send a male astronaut and a female astronaut to the Moon in 2024. But with the changes of administrations in the United States, the public is skeptical about the agenda. The program has experienced delays and undergone a major restructuring with the first human landing targeted for 2028. Two missions have been launched as of the publication date of the book. Artemis I (16 November–11 December 2022) was the first uncrewed integrated flight test of the Space Launch System (SLS) rocket and Orion spacecraft, while Artemis II (1 April-10 April 2026) was the first crewed mission of the program to orbit the Moon.

What is the relationship between government-funded manned space exploration missions and space tourism? As mentioned earlier, the advantage of government programs is that government can pool all the necessary resources together and motivate stakeholders to make it happen, but this is not sustainable. When the urgency of the political pursuit outweighs the cost of investment, the program obtains the green light and is adopted for implementation. Space tourism, on the other hand, is a self-sustaining market-oriented business that follows its own development path. Despite this, they are intrinsically intertwined.

Firstly, scientific discoveries from the government-funded manned exploration program can provide a great deal of scientific and environmental knowledge for the commercial space program, laying a solid foundation for space tourism and reducing the risks.

Secondly, during the period of suspension for the government program, the relevant technology will be transferred to the space tourism industry, hence reducing the cost of research and development for the commercial program and increasing its feasibility and safety.

Thirdly, if the commercial program is ahead of the government program, the government can become a user of low-cost space transportation, fuel extraction and refueling, space hotels, and lunar hotels. This kind of reliable users could further facilitate the development of commercial programs, while reducing the research and development costs for government programs.

Therefore, government-funded manned space exploration program and space tourism are mutually supportive. No matter which one becomes the forerunner, it will facilitate and support the other.

References

1. R. Poole, *Earthrise: How Man First Saw the Earth* (Yale University Press, New Haven, 2008)
2. D.H. Meadows, D.L. Meadows, J. Randers, W.W. Behrens III, *The Limits to Growth: A Report for the Club of Rome's Project on the Predicament of Mankind* (Universe Books, New York, 1972)
3. R.B. Fuller, *Operating Manual for Spaceship Earth* (Lars Muller Publishers, Zurich, 2019)
4. C. Liu, *The Three-Body Problem* (K. Liu, Trans.). (Tor Books, 2014)

Manifold Space Exploration Vs Space Function

Chapter 4
New Space

Modern space endeavor has its roots in World War II and has experienced rapid development in the Cold War era. By the early twenty-first century, it had grown into a vital industrial sector with a market value exceeding hundreds of billions of dollars. Since the early 2000s, a transformative force has emerged to reshape the trajectory of space industrialization, which is dubbed "New Space". This new paradigm necessitated a corresponding term for the traditional, government-dominated approach: "Old Space". This section examines the distinctions between Old Space and New Space, as well as the role of space tourism in this context.

4.1 Old Space

Old Space refers to the philosophies and methodologies of systems engineering developed over the years to implement government-dominated space missions. Besides the aspect of technologies, it also encompasses project scheduling, cost management, and quality control principles, which are largely shaped by the imperative to achieve political objectives. For instance, China's space sector upholds and operates under the principles articulated by Premier Zhou Enlai in the mid-1960s for major national defense projects, i.e., Serious and Diligent (*Yansu Renzhen*), Meticulous and Thorough (*Zhoudao Xizhi*), Steady and Reliable (*Wentuo Kekao*), Flawless and Perfect (*Wanwu Yishi*). These principles remain a guiding doctrine for space agencies and enterprises with an emphasis on absolute mission assurance.

Moreover, on the basis of self-reliance and guided by the systems engineering principles of Qian Xuesen,[1] China's space program has developed a distinctive project management methodology. For the organizational architecture, there must be a

[1] Qian Xuesen, 1911–2009, the father of Chinese Rocketry who synthesized Soviet missile technology with systems engineering.

J. Wu, *Space Tourism*, SpringerBriefs in Applied Sciences and Technology,
https://doi.org/10.1007/978-981-92-0883-8_4

Systems Integration Department, a centralized unit responsible for top-level design and cross-system coordination, which is unique to China's aerospace structure. For the functional segmentation, a mission is decomposed into Launch Vehicle System, Satellite System, Tracking, Telemetry and Command System, Launch Site System, and Application System. Every system is further divided into functional subsystems, and granularity can be added down to the next level.

For the management, the Dual-Track Management is adapted from the Soviet space program, i.e., the administrative line and the technical line. In the administrative line, Chief Commander, System Commander, and Subsystem Commanders are appointed. In the technical line, Chief Designer, System Designer, Subsystem Designers, and payload designers are appointed. The administrative line is responsible for the appointment, scheduling, and quality management, while the technical line is responsible for the technical issues in the mission design and development.

For the development phase, it is roughly divided into three phases: the design phase (the prototype development phase), the validation phase (engineering model phase), and the production phase (flight model phase). These three phases are followed by testing at the launch site, launch campaign, in-orbit commissioning, handover, etc.

For the cost management, the cost evolution has transitioned from "cost-no-object" to full-cost accounting to guarantee quality and mission assurance, though costs remain secondary to political imperative. It is understandable that the cost remains high in Old Space.

For quality control management, the Dual-Closure System is adopted to make sure both the technical closure and the management closure are implemented when malfunctions arise, so as to avoid similar mistakes in the future. The standards of technical closure include Where, Why, Repetition, Solution, and Lessons learned. *Where* means to identify the accurate root causes of the problems. *Why* means to determine the physical and chemical mechanisms of the quality problems through tools such as theoretical analysis or tests. *Repetition* means to repeat the process that caused the quality problems through methods such as tests and other verification tools to verify the above-mentioned root causes and mechanism determination. *Solution* means to adopt effective measures to correct the quality problems and further ensure the effectiveness of the measures taken. *Lessons learned* refers to sharing the quality problems with all the entities involved in mission model development and beyond.

The standards of management closure include How, Who, Solution, Responsibility, and Regulation improvement. *How* means to examine the whole process where problems arise and develop, to find out the weak link or management loopholes. *Who* means to identify the accountable entities and persons. *Solution* means to work out and adopt corrective and preventive measures against the weak link and management loopholes. *Responsibility* means to treat seriously the accountable person in management errors so as to improve management. *Regulation improvement* means to improve and exercise regulations against the weak link and management loopholes.

Any changes involving technical status need to pass the review against five standards, i.e., Study, Consensus, Verification, Approval, and Implementation. *Study* means to carry out a thorough study of the necessity for the change, the technology replacement, and the implementation scheme. *Consensus* means that the implementation scheme needs to be confirmed by all the involved systems, subsystems, and the smaller subsystems. *Verification* means to verify the implementation scheme through tests, to prove the changes are effective and implementable. *Approval* means that changes need to be approved by a higher-rank person in charge of the subsystems/systems. *Implementation* means that all the changes which pass the review and get approved should be executed according to the technical implementation scheme, and relevant technical documents should be updated accordingly.

To guarantee absolute mission success, Chinese aerospace engineers have formulated unique quality control protocols. Most notably, the Nine-Risk Management System specifically addresses potential failures arising from new technologies, new materials, new processes, new configurations, new environments, new organizations, new positions, new personnel and new facilities. This implies that anything new is not good, and to make sure of absolute success, anything new should be reconsidered with extreme caution [1].

In short, the design and management measures developed by the Old Space have been very effective in ensuring the complete success of missions, and have ensured brilliant achievements of China's space industry over the past 60 years. At the same time, however, its industrial system is plagued with a heavy burden, resulting in high development costs.

4.2 New Space

In the 1970s, satellite communications began to take hold of the civilian commercial market, and these satellite missions have been primarily driven by market demand ever since. This marked the first break from the single-user model where all space missions were government-funded. However, the development of communication satellites was still undertaken by institutions or enterprises with long-standing government space mission contracts.

In the 1980s, a professor at the University of Surrey in the UK began investigating the independent development of low-cost small satellites. He selected commercial off-the-shelf (COTS) components and tested them onboard satellites, and consequently developed small satellites under 100 kg at lower cost. Later, he further miniaturized these satellites and offered satellites under 10 kg. From a satellite development perspective, this breakthrough challenged the entrenched notion that high quality could only be achieved at high cost.

In the 1990s, a Harvard University professor proposed the CubeSat concept—a complete satellite integrated into a 10 cm × 10 cm × 10 cm cube, which has standardized CubeSats and further reduced costs. This inspired numerous companies to devote themselves to low-cost satellite development.

The early twenty-first century has witnessed Elon Musk's venture into launch vehicle development. He developed reusable commercial rockets, which significantly advanced the commercial space. The concept of New Space thus gained widespread acceptance [2].

The New Space refers to a series of systems engineering concepts and methods developed for commercial users, which prioritize the users' requirements. It encompasses reshaping the mission development process, the balance of low cost and quality at the same time, the overall management of the mission life cycle, and other new principles. These new concepts and management methods are in sharp contrast to the Old Space.

First, New Space missions are not government-driven but market-driven. This demand is not static, but evolves with economic and social development, as well as changes in people's living standards. Thus, New Space must continuously assess the market and adapt to the demands of the market. Moreover, as New Space operates in a competitive market economy, those who can accurately predict market trends and anticipate demands will capture the largest market share.

Second, leveraging existing technological foundations, New Space missions no longer require the establishment of a complete set of five systems (like Old Space), hence avoiding redundancy and reducing costs. Instead, New Space emphasizes meticulous Work Breakdown Structures (WBS) for each mission. The clearer the breakdown of tasks, the more manageable the systems engineering.

In R&D management, New Space abandons the hierarchies of separate administrative and technical lines. It typically adopts a project manager model where one individual oversees both. Notably, New Space favors flatter organizational structures, empowering frontline designers with greater authority and responsibility, including cost-conscious design. Technical and economic factors are integrated into the design process.

In terms of mission management, each mission is divided into six phases:

- Concept design phase includes but not limited to market demand analysis, problem definition, solution feasibility studies, adoption of new technologies, and identification of knowledge gaps.
- Mission development phase includes WBS, cost analysis, team allocation and labor distribution, new knowledge validation, as well as Assembly Integration and Testing (AIT).
- Production phase includes launch campaign and testing.
- Operational phase.
- Technical support phase.
- Mission termination phase.

The key difference from Old Space lies in bold application of novel knowledge and innovative technologies to cut costs and enhance performance. To mitigate associated risks, New Space relies on extensive experimentation and learns through trial-and-error, aiming to uncover and resolve all issues during mission development.

In summary, New Space is characterized by demand-driven pursuit, cost-consciousness, breaking conventions and formalities, lean structure, focus on systems level, encouraging innovation and bold experimentation, and streamlined processes. It aims at optimizing user-oriented performance and cost-effectiveness.

4.3 Space Tourism Is New Space

The section aims to answer what the relationship is between space tourism and New Space?

In terms of mission origin, space tourism is not a government program with direct government funding. It is inherently commercial, driven by growing tourist demand. Thus, space tourism unequivocally belongs to New Space.

In terms of cost control, space tourism relentlessly pursues reusability to reduce costs, aiming to bring the cost of spaceflight down to levels affordable for ordinary tourists. Therefore, space tourism is New Space.

In terms of risk management, space tourism does not aim for 100% success, but rather optimal safety under cost constraints. It mitigates risks through market mechanisms, such as insurance. This parallels the early days of commercial aviation—despite numerous plane crashes, the industry made solid progress through failures.

Therefore, space tourism practitioners must clearly recognize the uniqueness of their industry, and understand that as part of New Space it cannot rigidly follow traditional space management models. Instead, New Space principles must be applied to drive the development of space tourism to cater to the needs of customers and follow the laws of the market. In this way, the sustainability of space tourism can be ensured. This reflection also stems from the 50-year suspension of manned lunar missions since the Apollo Program.

References

1. J. Wu, *Introduction to Space Science* (Springer, 2021)
2. E. Berger, *Liftoff: Elon Musk and the Desperate Early Days that Launched SpaceX* (William Morrow, an imprint of HarperCollinsPublishers, 2023)

Chapter 5
Beyond Confines of the Earth's Gravity

As mentioned in Chap. 1, the microgravity environment is not the unique feature of space tourism. It is merely an artificial condition during orbital flight resulting from the equilibrium between centrifugal force and centripetal force (Earth's gravity). However, to escape the Earth's gravity to embark on a space tourism trip, we must first experience this environment by staying in Low Earth Orbit (LEO) prior to reentry or accelerating toward more distant destinations. Correspondingly, experiencing the microgravity environment in LEO becomes the first attraction of space tourism.

If the experience of sustained microgravity—or even brief airborne floating—is the goal of space tourism, we can use a series of methods on the ground to partially replicate these effects. Therefore, before introducing technologies to escape the Earth's gravity, we will first introduce several simulation methods for microgravity or reduced gravity environments, and then present technical solutions to overcome gravitational constraints.

5.1 Ground-Based Simulation Facility

If we consider entering the Earth's orbit and achieving sustained flight as the first true milestone of space tourism, there are still methods to attract potential tourists, which can also serve as pre-spaceflight training programs. These include ground-based microgravity simulation facilities and high-altitude free-fall experiences, such as parabolic flights and skydiving.

In reality, sustained microgravity simulation on Earth is extremely challenging. However, if we allow approximations—where conditions partially resemble the orbital microgravity environment—the following methods can be considered: wind tunnel floating, water tank buoyancy, magnetic suspension, and wire-suspended capsules. The following sections introduce the working principles of these methods.

J. Wu, *Space Tourism*, SpringerBriefs in Applied Sciences and Technology,
https://doi.org/10.1007/978-981-92-0883-8_5

1. **Wind Tunnel Floating.**

   This simulation method is quite popular for the time being, with facilities commonly found in amusement parks and even shopping malls. As shown in Fig. 5.1, the wind tunnel generates a powerful upward airflow to lift participants into suspension. While floating, participants can feel strong air currents, which is quite different from the true weightlessness experienced at the ISS. Yet, to observers outside the tunnel, the effect convincingly mimics the floating in space. Thus, it's less about the participant's experience and more about the spectator's illusion of floating in space.
2. **Water Tank Buoyancy.**

   According to Archimedes' principle, the buoyancy force exerted by any fluid on an object equals the weight of the fluid displaced by the object. Therefore, for a participant to achieve neutral buoyancy in water—where they neither sink nor float—they must wear counterweights to balance their body weight and the additional mass of their simulated spacesuit.

   Since the mass of displaced water from a person wearing an aquatic spacesuit simulator typically exceeds the combined weight of the person and suit, precise weighing prior to setting foot in the water is required to determine the necessary counterweights. Observers can monitor the participant through underwater viewing windows. For the participant, this environment closely resembles microgravity in space, though imperfect weight distribution, e.g., slight arm drift due to unadjusted limb buoyancy, creates minor deviations. Nevertheless, the overall experience is remarkably similar to that in space conditions, making this method

Fig. 5.1 Wind tunnel floating

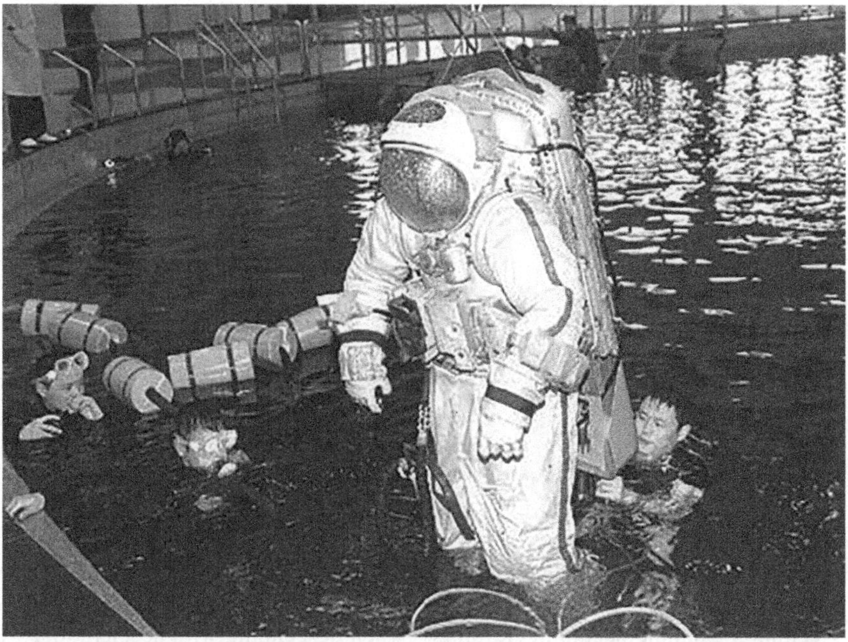

Fig. 5.2 Chinese taikonauts preparing for neutral buoyancy training

a cornerstone of astronaut training on the Earth. By wearing full mock-up extravehicular activity (EVA) spacesuits (including gloves), astronauts can accurately simulate extravehicular activities (EVAs). Figure 5.2 shows the neutral buoyancy training.

3. **Magnetic Suspension.**

   In principle, magnetic suspension faces no theoretical barriers. A straightforward implementation of this method involves installing a large magnetic field (size dependent on budget) above the participant, who wears a simulated spacesuit embedded with ferromagnetic materials. This setup can partially counteract the Earth's gravity to achieve suspension.

   The configuration of ferromagnetic components in the suit must be adjusted based on the participant's weight and the target gravity to be simulated—e.g., 1/6 Earth's gravity for lunar conditions or 1/3 Earth's gravity for Mars. However, similar to water tank buoyancy training, precise balancing for limbs remains impractical. Thus, participants may perceive discrepancies in gravity levels between arms and the body. Figure 5.3 shows the schematic of magnetic suspension.

   A key limitation of this method is the requirement for prolonged exposure to a strong static magnetic field. Current scientific findings suggest that brief exposure to such fields poses no health risks. Nevertheless, individuals with magnetically sensitive implants (e.g., pacemakers) or embedded metal objects must be excluded from this experiment.

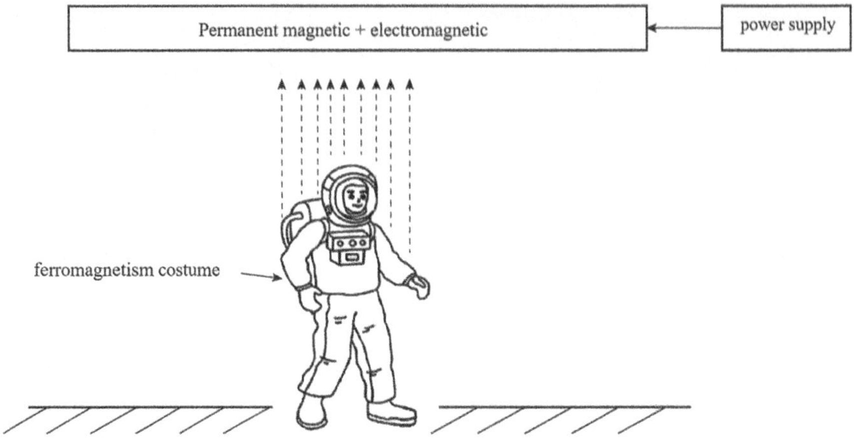

Fig. 5.3 Schematic of magnetic suspension

4. **Wire-Suspended Capsules.**

   While simulating zero-gravity environments on the Earth is extremely challenging, creating reduced gravity conditions is relatively straightforward. A common example occurs in elevators—when descending with acceleration, we experience a state of reduced gravity. This principle has been applied to simulate lunar gravity (1/6 Earth's gravity) and Martian gravity (1/3 Earth's gravity).

   Figure 5.4 shows China's comprehensive test site for extraterrestrial landing. The lander is suspended by cables and descends at predetermined acceleration rates to simulate target gravity conditions. During descent, the cable tension is dynamically adjusted to cooperate with the lander's reverse thrusters to accurately replicate gravitational forces, allowing observation of the lander's performance at touchdown.

By controlling the capsule's descent acceleration from its peak height, temporarily sustained reduced gravity conditions can be achieved. Depending on the facility height, the reduced gravity conditions typically last from several seconds to about 10 s. The capsule must be restored to the starting position after each test. The primary drawbacks of this method are the high construction costs and short simulation durations. However, its advantages are prominent: adjustable gravity levels (suitable for both lunar and Martian simulations) and repeatable test conditions.

Currently, China's only operational wire-suspended test facility is operated by the China Aerospace Science and Technology Corporation (CASC) for national missions. Interestingly, such facilities could also be constructed in commercial entertainment parks, serving dual purposes for tourism and scientific experimentation.

Fig. 5.4 China's comprehensive test site for extraterrestrial landing

5.2 High-Altitude Skydiving and Parabolic Flight

For off-ground experiences, methods like bungee jumping, skydiving, and parabolic flight aircraft can simulate gravity variations through free-fall from high altitude.

Bungee jumping provides seconds of weightlessness, but like skydiving, participants must overcome the psychological stress of leaping downward—this kind of stress is absent in actual space tourism.

Skydiving imposes greater psychological demands on the participants than bungee jumping due to the extremely high altitude (typically 4000+ m, while 50–100 m for bungee), despite the fact that there is an experienced instructor to dive with the participants. Both methods are plagued with the wind resistance issue, which is common in ground-based wind tunnels, impeding genuine free-floating sensations.

In contrast, parabolic flights deliver a far more authentic space-like experience. As illustrated in Fig. 5.5, the aircraft follows a parabolic trajectory. When the engine stops, the aircraft enters a free-fall, mimicking the ballistic curve of a thrown object. Passengers experience sudden loss of gravity, very similar to weightlessness in space, with no visual reference to the outside. Depending on flight altitude, each parabola provides 20–30 s of weightlessness. The surrounding reference points show no significant changes except for slight variations in the cabin's inclination, making it impossible for the participants to realize that the aircraft is actually in free-fall.

At this moment, all objects previously secured in the cabin start floating alongside the participants. Depending on the aircraft's altitude, this weightless state

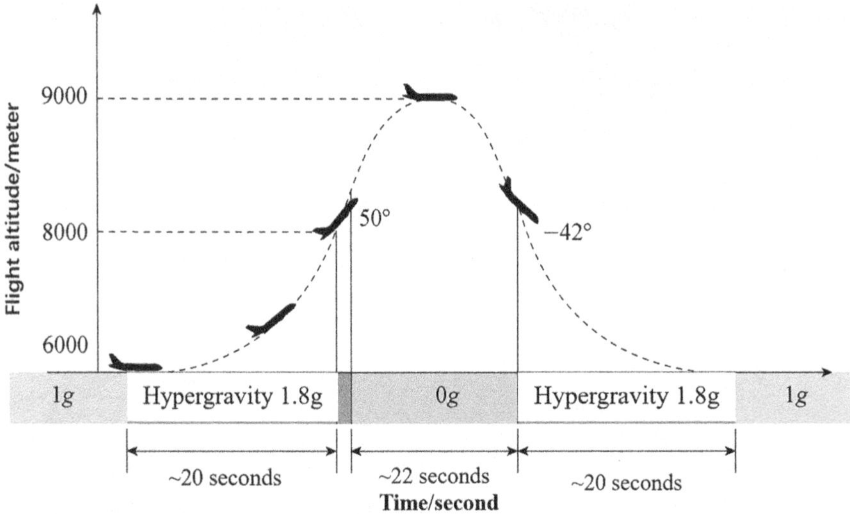

Fig. 5.5 Schematics of parabolic flight

typically lasts 20–30 s. As the aircraft dives to a predetermined altitude and must re-enter the powered flight period to climb up, participants will feel that gravity is returning, which is followed by hypergravity significantly exceeding normal Earth gravity (up to 1.8G). This cycle can be repeated several times or even a dozen times during a single flight.

Since participants need no spacesuit simulators and the conditions nearly replicate those aboard the ISS, parabolic flights are a primary tool for astronaut training. However, the cost far exceeds that of skydiving or bungee jumping [1].

5.3 Near-Space Flight

If not reaching the first cosmic velocity, a spacecraft can also ascend to very high altitudes with an extremely thin atmosphere—reaching 50 km or even 90 km, or crossing the Kármán line (100 km). After temporarily maintaining this altitude, atmospheric drag and the Earth's gravity will gradually reduce its speed and it will eventually return to the surface of the Earth. This complete process constitutes near-space flight [2]. Transporting tourists in this way is called near-space tourism. Even though it doesn't achieve orbital velocity and stay in Low Earth Orbit (LEO) in a sustainable manner, it is still a part of space tourism.

At altitudes above 50 km, where atmospheric density drops below 1%, the color of the sky transitions gradually from blue to black. Passengers can observe the Earth's curvature and thin atmospheric layers. However, reaching such altitude poses much higher demands on the propulsion systems than those of conventional aircraft, and the working principles are fundamentally different. Within the low atmosphere, engines primarily generate thrust by accelerating incoming air. However, above the stratosphere (>20 km), air engines become inefficient, necessitating replacement by scramjet engines or liquid-fuel rocket engines.

At these altitudes, rocket engines, which rely on reaction forces from expelled mass, demonstrate significantly higher efficiency than air-breathing alternatives like fuel-combustion aircraft engines. However, rocket propulsion alone is not suitable for the flight after reentry. Consequently, near-space vehicles typically employ hybrid propulsion systems.

Another factor to consider is that within the atmosphere, flight generates lift force proportional to the square of velocity, which is why wings are necessary for airplanes. In space, however, the absence of air means lift force cannot be generated through speed, which explains why space stations and satellites don't have wings. What may appear as wings are actually solar panels. Due to their wing-like appearance, solar panels are sometimes humorously referred to as "solar wings."

Near-space flight presents a unique scenario: it requires a hybrid between atmospheric flight and orbital flight. At altitudes of several dozen kilometers, the air becomes extremely thin, but if the vehicle moves fast enough, limited lift can still be generated. This is why near-space aircraft, operating between the atmosphere and space, typically feature small wings or deformable wings. At lower altitudes and slower speeds, the wings are completely extended; as altitude increases, air density drops, and speed rises, the wings retract or shrink in size.

To mitigate the effects of atmospheric variations at different altitudes by comprehensively utilizing flight parameters like velocity and lift, a hybrid flight solution has emerged. Virgin Galactic's near-space crewed vehicle adopts a two-stage design: a carrier aircraft (mothership) that transports the spaceship to 10,000–20,000 m high, and the spaceship that then separates from the mothership to fly higher. The large-winged carrier aircraft autonomously returns to the ground after separation, while the smaller-winged spacecraft uses rocket propulsion to reach higher near-space altitudes, which is followed by controlled gliding descent.

This two-stage approach represents the most economical method for near-space tourism, featuring full reusability of both vehicles. However, it cannot achieve the first cosmic velocity required for orbital flight, thus not qualifying as true space tourism. Virgin Galactic's implementation of this approach has faced multiple schedule delays. On July 11, 2021, it had its first successful commercial flight, with a ticket price of $250,000 per person. This pricing targets high-net-worth individuals within upper-middle-income brackets, establishing a sustainable market segment. The primary limitation remains: the weightlessness duration is too short, with most mission time devoted to ascent and reentry.

Also targeting near-space tourism, Blue Origin took an even simpler approach to capitalize the market opportunity. They designed a total vertical launch with a fully reusable rocket engine, and the passenger capsule re-enters with the help of a parachute. All designs use the most low-cost and mature technologies, but can still reach an altitude of 110 km. On July 20, 2021, they also achieved success in their first commercial flight just 9 days after Virgin Galactic's, becoming a strong competitor for Virgin Galactic in this market.

Distinctions between near-space travel, LEO tourism, and lunar tourism will be elaborated on in the following chapters.

5.4 Access to LEO

1. **Vertical Launch.**

   Since Konstantin Tsiolkovsky, the founding father of astronautics, the multi-stage liquid-fuel rocket with vertical takeoff has been regarded as the optimal solution for escaping the Earth's gravity.

   First, liquid propellants offer the highest energy density, enabling rocket engines to achieve maximum specific impulse. In other words, liquid fuels generate more thrust per unit mass than solid propellants.

   Second, multi-stage rockets discard empty fuel tanks and engines during flight, hence reducing weight and improving overall efficiency.

   Finally, atmospheric drag increases exponentially with velocity. Vertical launch allows rockets to swiftly leave behind the dense lower atmosphere at low speeds. When the rocket gains the momentum to speed up, air drag becomes significantly smaller in the thinner upper layers. From Robert H. Goddard's first liquid-fueled rocket to Wernher von Braun's V-2 rocket, spaceflight has followed this trajectory. With advancements in propellant chemistry and reduced mass in rocket structures, the mass of the payload to be sent to space is becoming larger and larger. Despite that, the fundamental method of escaping the Earth's gravity remains unchanged.

   To reduce costs, several revolutionary changes have occurred. The first major shift involved engine configuration. It is the normal practice to increase engines by stage for multi-stage rockets. To accommodate the different requirements of payloads, strap-on boosters were later adopted. For instance, a core stage with 4 engines plus 4 strap-on boosters means that 8 engines will be at work during liftoff. Compared with the cost of the structural frame, the cost of engines is the highest, making it necessary to shift from multiple smaller engines to a single high-thrust engine. To optimize payload flexibility, rideshare launches and piggyback payloads were introduced to fully utilize the engine's residual capacity. This transition, exemplified by the update from Ariane 4 to Ariane 5, reduced launch costs from $50,000/kg to ~$20,000/kg.

   The second revolutionary change was pioneered by SpaceX. By mass-producing smaller, low-cost engines and clustering them (9 engines per core stage: 8 active +1 redundant), SpaceX achieved economies of scale despite more engines involved.

   The next leap was core stage reusability. Since 2008, SpaceX has leveraged the progress of high-speed computers and AI technology to solve real-time challenges like unpredictable weather during landings, enabling precise recovery. With the increasing practical reuse rates (e.g., Falcon 9 boosters reused up to 29 times by July 2025), the launch costs have plummeted from $10,000/kg to $1000/kg.

2. **Horizontal Takeoff.**

   As previously mentioned, Virgin Galactic's SpaceShipTwo employs the horizontal takeoff method. The mothership carrying the spaceship takes off like a conventional airplane on a runway, utilizing aerodynamic lift to reach a higher altitude (approximately 15 km). Unmanned air-launch systems demonstrate larger capacity, with some air-launched rockets reaching 20 metric tons. However, the mothership's maximum speed (~0.3 Mach) is merely 10% of the velocity of the first stage, limiting its ability to replace conventional boosters. Consequently, air-launched vehicles retain significant dry mass. Even a 20-ton air-launched spaceship cannot achieve the first cosmic velocity. Thus, SpaceShipTwo only can be used for suborbital tourism, with peak altitudes below 100 km (the Kármán line).

3. **Inclined Takeoff.**

   Another attractive solution involves magnetic levitation (maglev) tracks for launching spacecraft into space to minimize friction. A mobile launch platform equipped with strap-on rocket boosters would transport the spacecraft to a 5000-m mountain summit before rocket ignition. This provides an initial velocity of ~50% of the first stage, allowing a smaller first stage to achieve the first cosmic velocity (7.9 km/s) after separation. The smaller first stage is recoverable and reusable. This approach ensures full reusability of all components of the rocket, while significantly improving launch capacity as against the horizontal takeoff, making it viable for crewed space tourism. The maglev platform and attached boosters remain on the ground track, eliminating complex recovery operations. The advantage of this approach is nearly 100% reusability, while the disadvantages include high initial construction cost for infrastructure and fixed orbital inclination, which makes it cost-effective only for high-cadence launches such as daily tourist flights and Sun-synchronous orbit missions.

Among these three approaches to escape the Earth's gravity, vertical launch remains the most mature, with costs still declining. Currently, it is the most viable solution for sustainable space tourism. Note that atmospheric reentry challenges and costs will be analyzed separately in the following chapters.

The horizontal takeoff method strikes a balance between the launch cost and the reentry cost. However, its primary limitation lies in its launch capacity, i.e., the limited payload capacity and orbital velocity. These challenges are fundamentally linked. If these technical hurdles are surmounted, its integrated launch-and-recovery capability could slash overall costs. The key innovation required is a dual-mode engine, e.g., a counter-rotating scramjet, that operates efficiently both in high-density and low-density atmospheres.

Inclined takeoff remains theoretical and needs further design and study. It requires R&D funding for test track construction. Space tourism is not the obligation of the government, so it requires commercial financing to secure sufficient construction funds.

A space elevator is another way to launch, which relies more on infrastructure and related technologies. It is frequently depicted in science fiction. Although space elevators are theoretically possible, their practical implementation is still decades away. Therefore, it is not considered in this analysis.

To sum up, rockets with vertical launch are still the mainstay to break free of the Earth's gravity. However, it remains unclear whether using multiple low-cost engines in parallel or a single high-thrust engine is more advantageous. This depends, to some extent, on both cost and safety concerns. For near-space tourism, the technical and economic feasibility of horizontal takeoff has been validated, suggesting that this approach is likely to continue developing in the foreseeable future. The inclined takeoff approach, combining magnetic levitation with rocket engine boosters, remains in the theoretical stage. Whether it can replace vertical launch in the future depends primarily on its overall technical and economic feasibility. But in view of the practical demand of space tourism for frequent and large-scale launches, the inclined takeoff may be the most efficient and cost-effective way to enter space.

5.5 Reentry and Landing

To enter space, a spacecraft must overcome the Earth's gravity and reach the first cosmic velocity (~7.9 km/s) to achieve sustained orbital flight. Conversely, during reentry, a manned spacecraft must decelerate to gradually reenter the atmosphere to reach the Earth's surface. Accelerating a spacecraft to the first cosmic velocity is challenging, while decelerating it from the first cosmic velocity or even the second cosmic velocity (~11.2 km/s) is equally difficult. However, due to the Earth's gravity, any spacecraft traveling at a speed below the first cosmic velocity will eventually return to the Earth. The primary engineering challenge is to ensure a rapid and safe return. For space tourism, cost-effectiveness must also be considered.

Deceleration requires fuel and the ignition of retro-rocket engines, which must be launched from Earth, thus inevitably increasing the launch mass. Ideally, if we assume the ability to reverse the launch process, i.e., the spacecraft decelerates and descends gradually in exact reverse sequence, the fuel required would somehow equal that used during launch. Therefore, in this case, we need to double the fuel for a round trip, further increasing launch weight. Such an approach is neither economical nor necessary. Thanks to the Earth's gravity, the spacecraft will inevitably descend. The real challenge arises when the spacecraft rotates 180°, activates its retro-rockets to reduce its orbital speed, and then starts a nearly free-fall descent. Like a meteor entering the atmosphere, the spacecraft experiences extreme aerodynamic heating generated by the friction between the atmosphere and the spacecraft, with temperatures exceeding 1600 °C. How to prevent this heat from damaging the spacecraft is the most critical engineering issue during reentry.

The engineering solution typically involves high-temperature-resistant ceramic coatings or ablative materials to dissipate heat by burning away. However, larger spacecraft have greater windward surfaces, i.e., areas exposed to atmospheric friction, requiring more extensive heat-shield coverage, which raises costs and increases risks. A comparison between the U.S. Space Shuttle with large wings for glided reentry and the reentry module of China's Shenzhou spaceship reveals that the Shuttle's larger heat-dissipating area posed significantly higher reentry risks than the Shenzhou's compact design.

For the reentry, during the period of intense atmospheric friction, at altitudes between 70 and 110 km, the air near the reentry module's ablative surface is ionized by extreme heat, forming a plasma sheath. This sheath stops electromagnetic wave propagation, resulting in approximately 10 min of radio blackout, during which the module loses communication with ground control. Fortunately, all trajectories are precalculated, and no critical actions are required during this period. If communication is urgently necessary, signals can be relayed via an antenna at the spacecraft's tail section to a geostationary relay satellite, bypassing the plasma sheath. The un-ionized atmosphere on the opposite side of the plasma layer still allows microwave transmission.

After passing the blackout zone, the manned spaceship's speed drops below 1 km/s, allowing further deceleration via parachutes or controlled gliding wings. The lift generated by the wings allows the vehicle to enter a controllable state, similar to that of an airplane. Landings assisted by spacecraft wings are far smoother and safer, but significantly increase costs. Parachute-assisted landings, though cheaper, demand larger landing zones (dozens to hundreds of square kilometers) due to unpredictable wind conditions after parachute deployment. This is because, once the main parachute is in full swing, it is no longer possible to precisely determine the landing point using ballistic calculations; instead, it largely depends on the weather conditions at that time, such as wind speed and direction. Unlike government missions, space tourism missions cannot delay the return trip for ideal weather, necessitating close coordination with ground rescue teams. Ground rescue personnel need to reach the landing site as quickly as possible to facilitate the passengers' moving out of the capsule and subsequent transportation. Before touchdown, an altimeter-triggered retro-rocket (or reverse thruster) ignites to cushion the impact.

Compared to land landings, splashdowns (Fig. 5.6b) pose fewer safety risks at the landing site but greater recovery challenges caused by bad weather. First, marine weather is unpredictable, with limited historical data. Second, rescue response is slower due to rescue ships' lower speed as compared with ground vehicles (despite helicopter support). Third, waterproofing requirements for the capsule should be considered. Impact on seawater is smaller, making retro-rockets unnecessary at splashdown.

However, not all countries possess the conditions for land-based recovery. Currently, only Russia and China employ land-based touchdown and retrieval, while the United States has consistently opted for oceanic splashdowns. This divergence stems from two key factors: (1) Geographical Constraints: The U.S. launches crewed missions primarily from Cape Canaveral, Florida, where trajectories immediately pass over the ocean. In mission abortion scenarios, emergency landings would inevitably occur at sea. (2) Powerful Naval Forces: The U.S. Navy maintains a global presence, enabling rapid deployment of recovery forces. The vast expanse of the ocean offers far larger and safer splashdown zones than the continent, justifying America's long-standing preference for splashdowns.

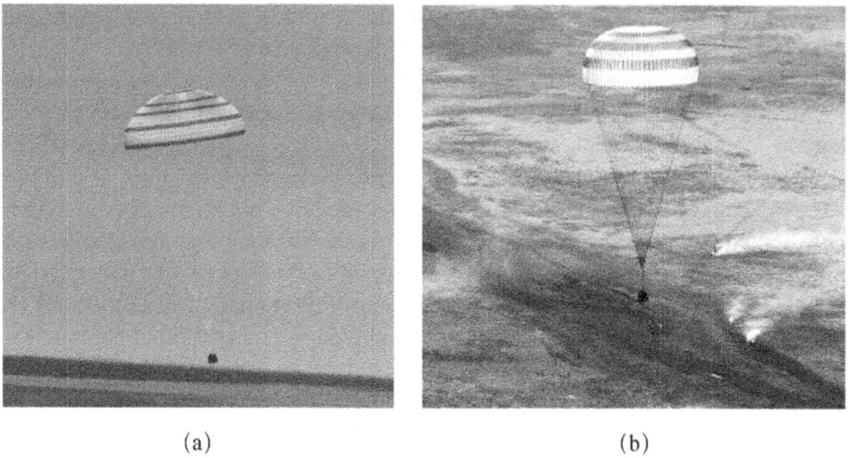

Fig. 5.6 Reentry module landing: (a) Land landing; (b) Splashdown

Fig. 5.7 Space Shuttle's horizontal landing

For winged spacecraft like the U.S. Space Shuttle or Virgin Galactic's SpaceShipTwo, atmospheric lift can decelerate the vehicle, making it possible to land on a designated runway. For space tourists, it is the optimal method for landing. Yet there is stark cost disparity between a blunt-body capsule and a winged spaceship, both in terms of construction and reusability. Therefore, as technology matures, the future landing method for space tourism missions will likely start with low-cost capsules and then transition to winged horizontal landings (Fig. 5.7).

References

1. V. Pletser (ed.), *Preparation of Space Experiments* (IntechOpen, 2020)
2. E. Seedhouse, *Suborbital: Industry at the Edge of Space* (Springer Praxis Books, 2014)

Chapter 6
Space Segment and Transportation Systems

Chapter 5 focuses on the methods to simulate microgravity on the Earth, the ways to overcome the Earth's gravity, and the approaches to enter space and return to the Earth. This chapter is devoted to the space segment and space transportation systems, including transfer stations like space stations or orbital hotels, as well as an Earth-Moon spaceship that travels between the Earth's orbit and the Moon. The primary reason to build separate Earth orbit transportation and space segments is to minimize costs. Overcoming the Earth's gravity requires enormous energy and investment, so infrastructure not involved in frequent return trips to the Earth can remain in space. In this way, the cost for repeated launches is saved.

For space tourism, the high-frequency transportation demand to and from the Earth orbit primarily comes from tourists [1]. For short-duration trips, an average round-trip payload capacity of 100 kg per passenger is generally sufficient. Other permanent space infrastructure can be launched and remain in orbit to reduce cost. This is particularly true for the travel between the Earth and the Moon.

6.1 Space Stations in the Earth Orbit

After completing their official missions, especially after completing critical scientific research, the International Space Station (ISS) and the China Space Station (CSS) can also be repurposed for commercial use if circumstances permit.

Current space stations, designed for scientific research purposes, are equipped with numerous scientific instruments and facilities in addition to essential operational systems. When converted for commercial use, these scientific instruments would be removed, leaving more interior space for commercial purposes. Meanwhile, with technological advancements, space stations will incorporate additional life support systems to reduce costs. These may include water recycling systems and in-house cultivation of plants and even animals to provide food for tourists. Most

J. Wu, *Space Tourism*, SpringerBriefs in Applied Sciences and Technology,
https://doi.org/10.1007/978-981-92-0883-8_6

importantly, the living areas and observation decks would be expanded to enhance the visitor experience.

Commercial companies could also build more specialized space hotels or Earth orbit hotels. Their designs would focus more on tourist experience, potentially using inflatable module designs to increase living space, including small private quarters similar to hotel rooms.

Orbital space stations function more like satellites, and are therefore typically developed and manufactured by satellite manufacturers rather than rocket manufacturers.

For crewed orbital space stations or space hotels dedicated to space tourism, the essential capabilities should include, but not be limited to:

- Capability for long-term orbital maintenance and attitude control over multiple years.
- Capability for orbital rendezvous and docking with crewed spaceship and cargo spacecraft.
- Life support with maximum water and air recycling for crew and tourists, by periodic unmanned cargo spacecraft resupply.
- Capability to support extravehicular activities (EVAs) by crew and passengers for necessary maintenance and spacewalk tours.
- Designated observation areas with excellent views of both the Earth and space.
- Facilities for exercise, entertainment, microgravity experience, and science education for tourists.
- Real-time communication (voice and video) with the ground.
- Microbial and viral detection and sterilization capabilities.
- Emergency escape systems for dangerous situations.

The long-envisioned space city would be a massive wheel-shaped structure generating Earth-like gravity through centrifugal force produced by rotation. Its orbital position should be distant from the Earth to capture panoramic views of our home planet and to gain a stable orbit, such as a circular equatorial orbit beyond geostationary orbit (GEO). Why not GEO? There are two reasons. First, the GEO slots are extremely limited and are primarily allocated for remote sensing and communication satellites. Second, a space city in GEO can only offer views of partial Earth. For instance, a station fixed above China can observe China, but not the USA, and vice versa. By placing the space city in a higher equatorial orbit, visitors could enjoy global views of the Earth in an orbital phasing cycle determined by altitude. Alternatively, positioning it at the Earth-Moon Lagrange Point L1, a gravitational equilibrium between the Earth and the Moon, would provide simultaneous close-up views of both the Earth and the Moon.

Given current technological capabilities and massive investment, constructing humongous wheel-shaped space cities like those depicted in science fiction remains elusive. The most practical and immediate priority is to construct orbital space hotels for space tourism in Low Earth Orbit (LEO). The construction of a rotating space city at the Earth-Moon L1 point only becomes feasible after achieving sustainable development of both LEO tourism and lunar tourism which relies on the Earth-Moon transportation systems.

6.2 Earth-Moon Spaceship

Following LEO space tourism, the Moon naturally emerges as humanity's next desired destination. Before achieving a lunar landing, the lunar orbital tourism mission would serve as the preliminary phase. Three transportation methods can facilitate the lunar orbital mission.

1. **Direct Earth-Moon Round Trip.**
   This approach requires launching all necessary fuel from the Earth in a single mission, demanding a massive rocket like the Saturn V used in the Apollo program. The immense launch cost makes this method economically unfeasible for commercial tourism.
2. **Earth Orbit Rendezvous with a Pre-parked Propulsion Module.**
   A propulsion module with propellant is launched first and parked in the Earth orbit. Then, a spaceship is launched to dock with the propulsion module. This staged approach divides the transportation system into multiple launches. The only additional work here is space docking. Since manned spaceships capable of completing round trips between the Earth and the Earth orbit can fully utilize current mature designs, their costs are lower than those of the direct method. However, due to the long time it takes to travel from the Earth to the Moon, which requires several days, tourists staying in the cramped spaceship have limited space and would feel uncomfortable, which necessitates the redesign of a tourist spaceship to fly directly to the Moon.
3. **Dedicated Earth-Moon Shuttle Spaceship (EMSS).**
   A dedicated Earth-Moon Shuttle Spaceship operates exclusively between the Earth orbit and the lunar orbit, without landing on either the Earth or the Moon. The shuttle spaceship is not responsible for the round-trip transportation between the ground and the Earth orbit. Tourists transfer at the space station, while the shuttle spaceship constantly remains in space, requiring only refueling and resupply between missions. This method maximizes reusability, minimizes maintenance, and optimizes passenger comfort with a spacious cabin and large observation windows.

The Earth-Moon Shuttle Spaceship (EMSS) is a reusable, long lifetime transport system shuttling between the Earth orbit and the lunar orbit. Its key features include a spacious interior for multi-day passengers, and large observation windows for panoramic views. For every round-trip, refueling is almost the sole major maintenance requirement. A velocity increment of 3.6 km/s is needed to transit from the Earth orbit (the orbital velocity equals the first cosmic velocity) to lunar orbit, consuming hundreds to thousands of kilograms of fuel per trip. These fuels, if supplied fromthe Earth, would be extremely expensive. However, if the fuel comes from space, such as from the lunar surface, the costs would be significantly reduced. The cost of launching unmanned spacecraft loaded with fuel from the Moon's surface is much lower than launching them from the Earth. Lunar-derived fuel drastically cuts expenses, as launching from the Moon consumes far less energy due to its weaker gravity. In-orbit refueling of the shuttle spaceship at lunar orbit further reduces operational costs.

The shuttle spaceship requires acceleration from approximately 7.6 km/s (Earth orbital velocity) to about 11.2 km/s, necessitating a velocity increment of 3.6 km/s. For a spacecraft weighing over 2000 kg and carrying 6–7 passengers, each round-trip consumes hundreds or even more than 1000 kg of fuel. The duration of this journey is crucial, because excessively long transit time would increase the burden on the life support system (increase in water and food supplies, etc.). Therefore, a 2-day or 3-day trip is optimal for space tourism.

If acceleration could be evenly distributed throughout the journey, e.g., maintaining a constant 0.1 m/s^2 (second squared) acceleration, the spaceship would reach the Earth escape velocity of 11.2 km/s in about 10 h. In order to be captured by the Moon, the spaceship needs a deceleration to 1.3 km/s, which would require approximately 28 h at the same acceleration rate. Thus, excluding orbital maneuvers and direction changes, the Earth-Moon single trip would take roughly 1.5 days under continuous 0.1 m/s^2 acceleration/deceleration, making the total 2- to 3-day round-trip journey arrangement reasonable.

The selected 0.1 m/s^2 acceleration serves multiple purposes: this acceleration is similar to the lunar surface acceleration, allowing future moon tourists to gradually acclimate to the lunar environment in advance; it provides sufficient pseudo-gravity for practical comforts, like keeping water in cups, which brings convenience to the tourists. Ensure that the seating is specially designed to be adjustable, aligning with the acceleration vector. See Fig. 6.1 for passenger seat orientation relative to thrust direction.

Each shuttle mission requires significant fuel replenishment. Transporting hundreds to over 1000 kg of fuel from the Earth's surface to LEO for refueling proves substantially more expensive than producing fuel on the lunar surface and transporting it to lunar orbit. Two reasons can explain this cost difference. First, the Earth's stronger gravity means that the cost to enter orbit is much higher. Second, lunar orbit refueling means the shuttle spaceship can depart the Planet Earth with partially filled tanks, reducing initial launch mass.

Therefore, the most efficient refueling approach for shuttle spaceships operating exclusively between the Earth orbit and the lunar orbit is lunar orbital refueling. However, this requires established capability for large-scale, automated fuel production on the Moon. Until such capacity is acquired, Earth orbit refueling remains the default option—still more economical than non-reusable lunar spaceship launched from the Earth, since the shuttle spaceship's main structure stays in space for reuse.

Unlike government-dominated manned lunar missions, a tourist-oriented shuttle spaceship has some additional requirements, including, but not limited to, the following: (1) Capability for sustained low-gravity acceleration; (2) Observation windows with comfortable, adjustable seating and cabin space for indoor movement; (3) Broadband real-time communication (audio/video) with the ground.

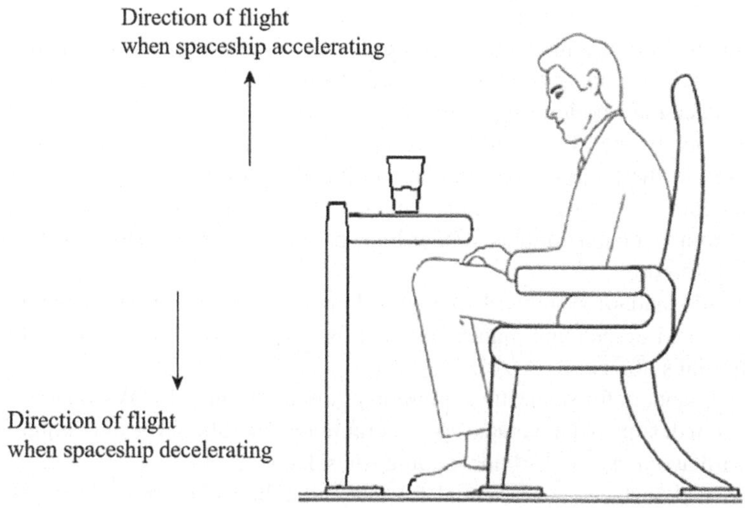

Fig. 6.1 Passenger seat orientation relative to the direction of flight

6.3 Space Station in Lunar Orbit

If humanity can establish space stations in the Earth orbit, then lunar orbit space stations are equally achievable. For space tourism, a lunar orbit station serves two primary functions: (1) acting as a transit hub for tourists, and (2) refueling Earth-Moon Shuttle Spaceship (EMSS) while receiving supplies from Earth orbit stations. Unlike research-oriented stations, tourist-oriented lunar stations eliminate scientific instruments, significantly expanding available interior space. Moreover, these stations can operate without the attendance of humans, allowing for more compact designs and reduced power requirements. During tourist visits, station operations would be managed by either the pilot of EMSS or the pilot of the Lunar Landing and Ascent Vehicle (LLAV).

Beyond standard docking ports for crew members (identical to Earth orbit stations), lunar orbit stations require one critical additional feature: the capacity to receive, store, and transfer lunar-sourced propellant extracted from the Moon's regolith to EMSS. All other functions mirror those of Earth orbit stations [2].

6.4 Lunar Landing and Ascent Vehicle (LLAV) and Its Operations

The Moon's escape velocity is 1/6 of the Earth's ($\approx$1.3 km/s to enter lunar orbit). With no lunar atmosphere, vehicles experience neither aerodynamic drag nor heating during high-speed flight, eliminating the need for thermal protection systems or streamlined fairings.

Parachutes cannot be used for deceleration in the airless lunar environment. The entire landing process must rely on powered descent using thrusters, making LLAV designs fundamentally different from Earth reentry vehicles, which is evident in the landers since the Apollo program.

Additionally, the LLAV will land on the Moon's surface shortly after picking up tourists from the lunar orbital station, and this duration typically does not exceed several hours. Passengers and pilots remain seated throughout this brief transit, permitting compact cabin layouts similar to a spaceship traveling between the ground and the Earth orbit.

Beyond standard systems, LLAVs need enlarged propellant tanks for powered ascending and descending maneuvers, and for transporting lunar-produced fuel to lunar orbital stations.

Unlike landers for scientific exploration, tourist-oriented LLAVs operate exclusively from designated lunar spaceports, eliminating additional design requirements for hazard avoidance and ad hoc landing site selection.

Hardened surfaces are essential for launch/landing platforms to mitigate dust contamination from exhaust-blown regolith during ascent/descent.

After landing, tourists will participate in various activities during the multi-day lunar stays, necessitating the construction of lunar hotels. At the lunar hotel, tourists will remove their spacesuits and enter a sealed environment. Tourists need to exit the LLAV in spacesuits, enter a sealed chamber for depressurization, and remove their spacesuits before accessing hotel areas. The airlock transition increases both energy costs and operational complexity. A more efficient approach involves landing the LLAV directly inside a cylindrical sealed chamber serving as a macro-airlock. The airlock chamber closes and inflates. Once it reaches 1 atmosphere of pressure, tourists can directly exit the spaceship and enter the hotel. The design of the cylindrical airlock can avoid the contamination caused by lunar dust particles being stirred up during landing and takeoff.

The Moon's airless environment means that ultra-precise trajectory calculations and landing point control are possible. Therefore, fixed-position airlock landing and ascending platform are technologically feasible.

Beyond passenger transport, LLAV performs three critical logistics functions, e.g., delivering supplies for staff on the Moon, transporting tourists to/from the lunar orbital station, transporting lunar-produced propellant to the lunar orbital station. Automated systems at the landing platform can directly connect to the fuel pipeline, enabling immediate refueling of the LLAV after the landing.

To optimize docking efficiency between LLAV and the lunar orbital station, the lunar hotel latitude (e.g., 45°) should match the lunar station's orbital inclination. This alignment minimizes propellant expenditure for orbital rendezvous and docking.

Figure 6.2 shows the schematic diagram of lunar tourism infrastructure.

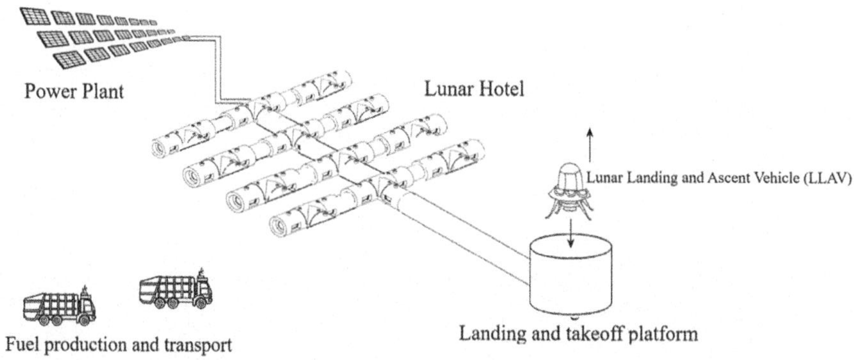

Fig. 6.2 Schematic diagram of lunar tourism infrastructure

References

1. M. Van Pelt, *Space Tourism: Adventures in Earth Orbit and Beyond* (Copernicus, Göttingen, 2005)
2. P. Stakem, *Lunar Orbital Platform—Gateway* (PRRB Publishing, Phillipsburg, 2018)

Chapter 7
Selection of Lunar Landing Zone

The most attractive parts of lunar tourism include rekindling the emotions of the Apollo astronauts when they landed on the Moon, experiencing the marvelous sensation of only 1/6 of the gravity of the Earth, and looking back at the blue Earth rotating alone in the dark universe. Unlike scientific exploration missions, lunar tourism does not change the landing zone every time. Therefore, what needs to be done only includes the selection of the most suitable location for tourists and the construction of tourism infrastructure. This chapter discusses the selection of landing zones and the corresponding problems associated with infrastructure construction.

7.1 Fuel Production

The selection of landing sites for lunar tourism missions needs to be considered with the problem of in situ fuel production. This is because the fuel production on the lunar surface is a prerequisite for reducing the cost of lunar tourism, and the produced fuel needs to be launched to a space station in lunar orbit for refueling the Earth-Moon Shuttle Spaceship (EMSS). The manned spaceship for the transportation between the lunar surface and the space station in lunar orbit can be used to transport fuel. Therefore, the landing site should be close to the site of fuel production, so that direct refueling of the lunar spaceship could be carried out once the fuel is produced.

The question is what is the best raw material to extract fuel from? I firmly believe that we should extract fuel from lunar regolith, not water ice.

Water ice as the raw material to extract fuel is quite popular. Why not water ice? There are two reasons: first, water ice is a limited resource that should be protected rather than consumed; second, it exists only at the bottom of craters where it is too deep to be reached by sunlight, and the rolling hills make it an unsuitable place for tourism.

J. Wu, *Space Tourism*, SpringerBriefs in Applied Sciences and Technology,
https://doi.org/10.1007/978-981-92-0883-8_7

Indeed, compared with the extraction of liquid hydrogen and liquid oxygen from the water ice, the extraction of these fuels from lunar regolith has an additional process, which involves extraction from other oxides, e.g., titanium oxide and iron oxide. The chemical reduction process involves heating the lunar regolith to over 800 °C, adding a small amount of hydrogen to combine the oxygen and hydrogen in the lunar regolith to produce water, and then ionizing the water to obtain the necessary oxygen and hydrogen. This, of course, requires additional energy. But it could be easily solved because during the lunar day there is abundant solar energy. In addition, in the lunar mare region, lunar regolith resources are in great reserves and are easily accessible. Therefore, in order to have easier access to lunar regolith for fuel extraction, we should choose lunar mares as landing sites for lunar tourism missions [1, 2].

The by-product of liquid hydrogen and liquid oxygen extraction is the lunar regolith slag, containing mainly titanium, iron, and silicon. If these slags are compressed into lunar regolith bricks, they will be very good building materials for the containment module or the lunar hotels on the lunar surface. Since the construction of the fuel extraction facilities comes first and the construction of the lunar hotel comes later, there is a constant supply of such building materials. Therefore, it is not necessary to build the lunar hotel underground or to build it in the shape of a bunker and cover it with lunar regolith.

7.2 Position of the Earth in the Sky

Looking back at the Earth from the Moon's surface, a faraway distance, brought about the second enlightenment from space, as discussed in Chap. 3. Leaving the Earth behind, the tourists have the opportunity to rediscover our home planet Earth. This is the highlight of lunar tourism. Therefore, the selection of the landing site for a lunar tourism mission should fully consider the convenience of looking back at the Earth to make sure the Earth appears in the right position in the sky.

Because of the tidal locking of the Earth to the Moon,[1] from the Earth we can only see one side of the Moon, and that side is always facing the Earth. The tidal locking affects how we see the Earth from the surface of the Moon—apart from a few degrees' libration movement, the Earth will stay roughly in the same position in the sky without Earthrise and Earth-set. But we can see the Earth spin in a roughly 24-hour cycle. In addition, about every 28 days, we will see a full illumination of the Earth (we can call it "full Earth" in line with the concept of "full moon") and a moment of total darkness when the Earth will be completely invisible.

[1] Earth's Moon rotates, but it takes precisely as long for the Moon to spin on its axis as it does to complete its monthly orbit around Earth. As a result, the Moon never turns its back to us, like a dancer circling—but always facing—its partner. This phenomenon, called "synchronous tidal locking," is actually quite common. All the solar system's large moons are tidally locked with their planets.

To determine the apparent position of the Earth in the lunar sky, we can use the origin of lunar coordinates (the point at the center of the Moon's near side) as the center. Then we can draw a circle with a radius extending to 15° lunar longitude or latitude. An observer at any landing site on this circle would see the Earth positioned 15° below the zenith (Fig. 7.1), corresponding to a 75° elevation angle above the lunar horizon. Similarly, if the circle's radius extends to 45° longitude or latitude, the Earth would appear 45° below the zenith (or at a 45° elevation angle above the lunar horizon) from any landing site on this circle, and so forth. Therefore, to ensure tourists enjoy a comfortable viewing angle when observing the Earth, the landing sites should be selected along a circle with a moderate radius—for instance, between 45° and 60°. Major lunar mares within this range include Mare Serenitatis in the northeastern quadrant, Mare Imbrium in the northwestern quadrant, Oceanus Procellarum (spanning the northwestern and southwestern quadrants), and Mare Nectaris in the southeastern quadrant.

Fig. 7.1 Lunar landing sites (on the circle) when the Earth appears 15° below the zenith

In addition to the comfort associated with the viewing elevation, the apparent orientation of the Earth must also be considered. If the landing site falls in the Moon's northern latitudes, the Earth will appear in the southern lunar sky, with its northern hemisphere at the top and its southern hemisphere at the bottom, which is consistent with the conventional orientation of the Earth maps. Conversely, if the landing site falls in the Moon's southern latitudes, the Earth will appear in the northern lunar sky, with its northern hemisphere at the bottom and its southern hemisphere at the top, which does not agree with our habit of viewing maps. Thus, landing sites in the northern latitudes offer a more familiar experience to view the Earth. If the landing site is on the lunar equator—for example, in Mare Serenitatis at 30°E, 0°N—the Earth will appear due west at an elevation of 60° above the lunar horizon.

However, if the landing site is close to the lunar south pole (as often discussed for exploration missions), the Earth would remain perpetually near the horizon, without rising up, and would appear upside-down with the Northern Hemisphere at the bottom. Moreover, since the Earth's diameter is 3.67 times that of the Moon, its apparent size as viewed from the lunar surface would be 3.67 times larger than a full Moon seen from the Earth. It would also display a striking blue hue with swirling white clouds. In commemoration of the 40th anniversary of the first manned landing on the Moon, Apollo 11 astronaut Michael Collins said:

> I really believe that if the political leaders of the world could see their planet from a distance of 100 000 miles, their outlook could be fundamentally changed. That all-important border would be invisible, that noisy argument silenced. The tiny globe would continue to turn, serenely ignoring its subdivisions, presenting a unified façade that would cry out for unified understanding, for homogeneous treatment. The Earth must become as it appears: blue and white, not capitalist or communist; blue and white, not rich or poor; blue and white, not envious or envied.[2]

7.3 Lunar Landscape

Besides observing the Earth, tourists would naturally want to experience the Moon's unique landscapes. Flat lunar mares are akin to the Earth's vast plains, without towering peaks, massive boulders, or steep canyons. If tourists only land on flat areas of lunar mares, they will lose the precious opportunity. Thus, while selecting landing zones in lunar mares with rich resources, where fine lunar regolith facilitates energy extraction, we should also consider the proximity to elevated terrain, such as mountain peaks. Such features are abundant along the outskirts of lunar mares, including the 80-km-wide Archimedes Crater in central Mare Imbrium; the 600-km-long Apennine Mountain Range to the east of Mare Imbrium, home to the 6.1-km Mons Huygens, the tallest peak on the Moon; the lunar Alps and the Alpine Valley to the

[2] Space News. [2009-07-15] https://spacenews.com/statement-from-apollo-11-astronaut-michael-collins/

north of Mare Imbrium. Imagine this scene: you, dressed in a spacesuit, are standing on the smooth expanse of a mare, with towering mountains rising behind you. Against the dark void of space, the Earth hangs at a comfortable position in the sky like a blue marble, appearing 3.67 times larger than the Moon when observed from the Earth. At this moment, the continents are faintly visible with the northern hemisphere at the top. What a breathtaking picture!

Therefore, for the best viewing experience, landing sites should be located near mare edges, and the Earth is positioned above in the sky. In addition, there should be high mountains in the backdrop.

From the perspective of architectural design, future lunar hotels must incorporate carefully oriented windows—every room should have a glass window facing the Earth.

7.4 Transportation on Lunar Surface

For tourists in spacesuits, their moonwalk can cover a very limited distance. However, with the help of lunar rovers as transportation tools, they can have long trips. During the Apollo 17 mission in December 1972, astronauts drove their rover over 30 km. Therefore, when selecting landing sites, we should consider how far a rover can travel. What's more, we should even make sure that the playground is big enough for faster vehicles. Landing sites should not be confined to areas with limited diameter within a lunar mare, such as the 80-kilometer-wide Archimedes Crater, as this would severely restrict tourists' travel distance on the Moon's surface.

Mechanized lunar transportation was realized long ago. For example, electric-powered rovers were applied in the manned lunar landing missions over 50 years ago. Thus, for future lunar tourism missions, this should not be a technical challenge at all. What's more, the electric vehicle and autonomous driving technologies are becoming more mature. However, there are a few new requirements for the rovers dedicated to lunar tourism:

- Improved rover capacity: Rovers can accommodate 6 people (1 driver + 5 passengers) or even larger groups (e.g., 8–10 passengers) to meet the demands of tour groups.
- Enhanced stability designs: Tracked or multi-legged rovers are required for improved stability when driving at a high speed, which enables traversal of steep slopes and rugged terrain while maintaining rapid mobility.
- Extended range: Rovers capable of traveling over 50 km or farther are required.

7.5 Energy Supply at Lunar Night

Any activity on the lunar surface must solve the problem associated with the long lunar night, which lasts for 14 Earth days. As a consequence of the Moon's 28-day rotation cycle, nearly all lunar locations (except certain high peaks in the polar region) will be shrouded in this prolonged darkness, which is very different from the Earth. During the lunar night, the surface temperature plummets below −150 °C, even reaching −170 °C. Sustaining tourism operations during this period is not just an economic challenge (affecting facility utilization of landing operations and hotels, etc.), but also a problem of tourists' experience. The lunar night offers a unique spectacle: the Earth is largely illuminated by the Sun, producing the most stunning images—especially during the "Full Earth" at mid-lunar-night, when it is fully sunlit.

It is a technical problem to guarantee energy supply during the lunar night. Potential solutions include nuclear power, wireless transmission via microwaves or lasers, and extracting water from lunar regolith for fuel cells.

- Nuclear power: Highly efficient but not ideal for tourism missions due to safety concerns. Radiation shielding requires heavy metals like lead. Only if lead can be easily extracted from lunar regolith, rather than transported to the Moon from the Earth, can nuclear power become an efficient energy solution. Otherwise, we have to resort to other solutions to ensure the safety of the tourists.
- Extracting water from lunar regolith for fuel cells. However, compared with direct utilization of solar power, it is highly inefficient to electrolyze water into hydrogen/oxygen and reconvert it for power. In addition, it is a process that requires big infrastructure to provide the necessary support. It is a waste of valuable fuel resources, if we reduce into water the available liquid hydrogen/oxygen, which is used in space transportation system.

Therefore, the best approach so far is the direct utilization of solar power [3].

But how to harness the solar power during the lunar night? In the future, humans will extensively develop the Moon and build a power grid connecting the day side and night side of the Moon. But before that, the optimal approach is to obtain solar power in the lunar orbit and transmit it in a wireless way to the lunar surface during the lunar night.

Research into using microwave wireless transmission to send solar energy from the Earth's orbit to the Earth has been ongoing for many years. The microwave rectenna efficiency on the Earth approaches 10%. The Earth has an atmosphere that blocks the transmission. Compared with direct utilization of solar power, it is still very inefficient. On the Moon, without the atmosphere, transmitting energy with lasers or direct light waves could bypass atmospheric losses.

The proposal for energy transmission using lasers entails directly channeling electrical energy from solar panels to a high-power laser, which is then targeted at the user's solar panels on the surface via a large-aperture optical telescope, ensuring a narrow beam angle. The drawbacks include low energy conversion efficiency, with a significant amount of energy lost during the processes of converting light to electricity and then back to light, and the high cost of large-aperture telescopes.

Another technology under research and development involves directly reflecting sunlight from orbit onto the user's solar array on the lunar surface, without the need for energy conversion from light to electricity and then from electricity to light, thus achieving the highest energy utilization efficiency. However, sunlight is not ideally parallel light. It will diverge after traveling thousands of kilometers and cannot be focused. The research objective is to design mirrors with focal lengths exceeding thousands of kilometers (corresponding to the orbit altitude), and to assemble multiple such small mirrors into a large mirror, each with the ability to track autonomously. Back on the lunar surface, the user can simply install an angle reflector to provide feedback information which can keep the mirror array in orbit locked to the user's position at all times. If this option is feasible, it will be the most promising and feasible energy solution during the lunar night. Such orbital solar reflectors, or orbital solar power stations, can serve multiple lunar users on a time-sharing basis, as long as the user has a passive optical angle reflector on the lunar surface.

7.6 Protection Strategies Against Radiation and Micrometeorite

Unlike the Earth, the Moon lacks a protective magnetic field. This allows solar wind particles to directly reach its surface. Measurements indicate that solar wind flux on the lunar surface is only slightly higher than that in Low Earth Orbit (LEO). With proper shielding provided by the building materials of sealed habitats, e.g., lunar hotels, background solar wind radiation poses negligible risks. Short-duration outdoor activities on the surface also require no special precautions under normal conditions.

During solar eruptions, particle radiation becomes a serious concern. However, since such space weather events are predictable, all outdoor activities can be canceled before the arrival of high-energy particles. Additionally, lunar habitats must include storm shelters—reinforced areas with thicker walls to shield against elevated radiation fluxes from cosmic rays or solar storms.

Interplanetary dust particles in the solar system (typically <1 mm in diameter) pose minimal risk due to the Moon's weak gravity as compared with the Earth, which reduces their impact momentum. Standard habitat exteriors can provide sufficient protection. For rarely occurring larger micrometeoroids, we can adopt designs similar to the International Space Station's compartmentalized airlock: pressure-breach alarms trigger immediate evacuation to safe zones, and affected sections are isolated for repairs.

Pre-existing lunar lava tubes can be utilized to serve as shelters against solar particle radiation and micrometeoroid radiation. However, since such natural formations are geographically limited, landing site selection should prioritize energy accessibility and scenic value.

References

1. D. Rapp, *Use of Extraterrestrial Resources for Human Space Missions to Moon or Mars*, 2nd edn. (Springer, 2018)
2. E. Seedhouse, *Life Support Systems for Humans in Space* (Springer, 2020)
3. J.-H. Zünkler, *Electrical Power Systems for Lunar and Space Exploration: Fundamentals and Design*, 1st edn. (Hanser Publications, 2025)

Chapter 8
Construction of Lunar Hotels

In Chap. 7, issues related to site selection for lunar tourism are addressed. This chapter will specifically discuss matters concerning the construction of lunar hotels. Our discussion remains grounded in lunar tourism considerations, evaluating both technical feasibility and economic viability.

8.1 Solar Power Stations

Solar energy represents the most cost-effective and efficient power source on the lunar surface. Therefore, lunar hotels require large-scale solar power stations to tackle their energy demands.

Solar panel production must take place onsite. Currently, robots capable of autonomously manufacturing solar arrays are under development, with lunar surface production trials expected to take place in the near future. The production methodology mirrors energy generation principles—extracting silicon and other elements (e.g., germanium) from lunar regolith to fabricate semiconductor cells locally, potentially yielding byproducts like liquid hydrogen and oxygen.

The installed capacity of solar power stations must account for: efficiency fluctuations when receiving reflected sunlight from orbital solar power stations during lunar nights, and daily sunlight availability constraints. Additionally, power stations must meet the power requirements for battery charging before lunar night. A feasible scale would be hundreds of kilowatts supported by several thousand square meters of solar arrays.

Assuming one automated robot can produce 1 m^2 of solar array in a lunar day (with production halted during lunar nights), ten robots could manufacture approximately 1500 m^2 annually. This economic factor proves critical in lunar hotel planning.

As noted earlier, these lunar surface stations can also receive reflected light from orbital solar installations. The only required modification would be installing

J. Wu, *Space Tourism*, SpringerBriefs in Applied Sciences and Technology,
https://doi.org/10.1007/978-981-92-0883-8_8

passive optical corner reflectors within solar arrays to provide tracking feedback to orbital power stations.

8.2 Building Materials and Hotel Layout

Whether for fuel extraction or solar panel production, the byproduct is lunar regolith slag, which is rich in various metals and mineral elements such as titanium and iron. Through high-temperature and high-pressure processing, this slag can be transformed into lunar bricks, the fundamental building material on the Moon.

Additionally, many institutions and companies are conducting research on the use of raw lunar regolith as 3D printing material to fabricate structural components—or even entire habitats. Some enterprises are also exploring methods to extract metallic materials directly from lunar regolith or asteroid minerals for use in future space city construction.

All of these could serve as building materials for lunar hotels. However, the most cost-effective and readily available option is lunar bricks made from slag byproducts of fuel extraction, solar panel manufacturing, and glass production. Standardized lunar bricks are ideal for robotic operations, and are friendly for transportation, storage, and construction.

Currently, there are no test data on lunar bricks' radiation shielding effectiveness, making it impossible to determine an equivalent aluminum shielding thickness. As a temporary assumption, if we scale up the required thickness several dozen times and consider the operational constraints of fuel-extraction robots, the optimal brick size could be defined as 370 mm × 240 mm × 115 mm.

In lunar hotel construction, lunar bricks provide primary external shielding against space radiation while also offering thermal insulation. For a pressurized lunar habitat, an internal airtight layer capable of withstanding Earth-like atmospheric pressure (1 atm) is also required. This layer could use flexible materials for inflatable space hotels in the Earth orbit, which are transported from the Earth rather than manufactured on the Moon. Thus, the basic structure of a lunar hotel consists of an internal flexible airtight layer and an external structure of lunar bricks.

To ensure safety against air leaks or other hazards, lunar hotels should adopt a compartmentalized design similar to space stations in Earth orbit, dividing the structure into multiple independently sealed modules. Each module should be of limited size but include a shared corridor and a private room (see Fig. 8.1).

Corridor walls should be equipped with airlock hatches, allowing connection to adjacent modules. Therefore, there are isolation airlock doors between each module inside the lunar hotel that are normally open but can be closed at any time. Each module's ventilation is independently supplied from the outside, enabling isolated operation if necessary.

To enhance the guest experience, each private room should feature a large viewing window, ideally facing the Earth.

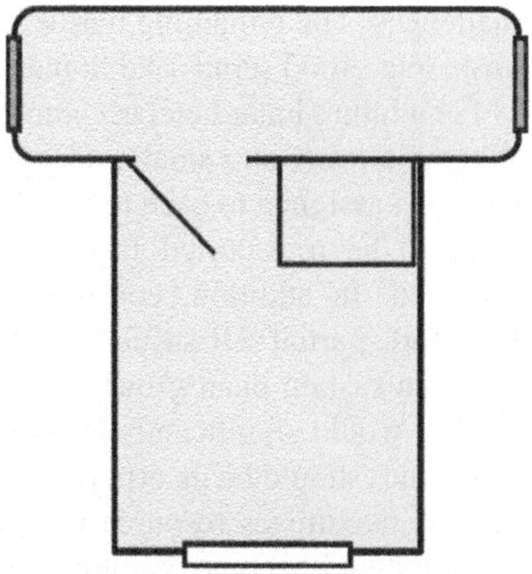

Fig. 8.1 A sealed module of a lunar hotel

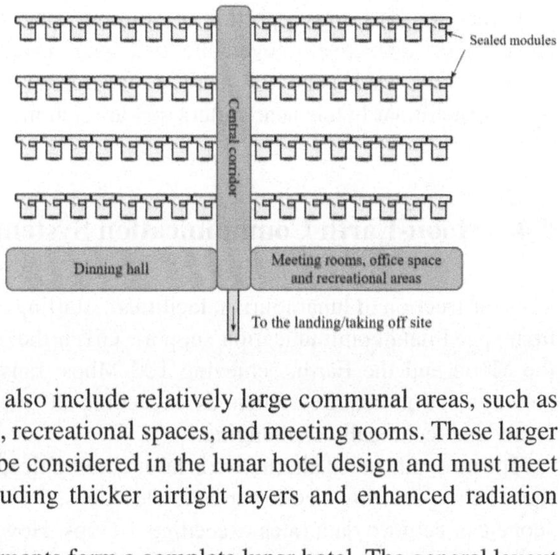

Fig. 8.2 Overall layout of a lunar hotel

A cozy lunar hotel should also include relatively large communal areas, such as central corridors, dining halls, recreational spaces, and meeting rooms. These larger non-residential areas should be considered in the lunar hotel design and must meet higher safety standards, including thicker airtight layers and enhanced radiation shielding.

When combined, these elements form a complete lunar hotel. The general layout includes a central main corridor, branch corridors with multiple private rooms, dining and conference facilities at the central corridor's end (see Fig. 8.2). Note that technical support facilities, such as power systems, water/air recycling, and life-support infrastructure, are not depicted in Fig. 8.2.

8.3 Water, Air, and Bioregenerative Life Support Systems

Given the requirements for energy efficiency, emission reduction, and environmental protection, water recycling technologies are already materialized on the Earth, while air revitalization systems have been successfully implemented on the International Space Station (ISS). Thus, these do not pose significant technical challenges. The remaining hurdle is bioregenerative recycling—converting organic waste (e.g., food scraps and human waste) into reusable resources.

For a future lunar hotel accommodating dozens to hundreds of occupants, the use of organic waste on a small scale to enrich lunar regolith for farming is crucial. Even 100 kg is assigned to each tourist as their upper mass limit; the amount of food and water to be transported from the Earth to the lunar surface remains severely restricted. In addition, consumption by hotel staff should also be considered. Therefore, partial self-sufficiency becomes essential. While current hydroponic systems can sustain plant growth using nutrient solutions, integrating enhanced lunar regolith would significantly improve recycling efficiency and scale. Thus, agricultural zones should be incorporated into pressurized modules where biowaste-treated regolith maximizes resource recovery. Growing fresh produce in enhanced lunar regolith could become a unique selling point for lunar tourism, which elevates the overall experience [1].

If each tourist brings a portion of drinking water to the Moon, and closed-loop recycling systems are employed, the total water reserves will accumulate over time. Consequently, water is not the scarcest resource on the Moon, eliminating the need to construct lunar hotels near craters that are rich in water ice.

8.4 Moon-Earth Communication System

The construction of lunar tourism facilities—starting from fuel extraction—requires high-speed data communication support. Given the 380,000 km distance between the Moon and the Earth, achieving 100 Mbps transmission rates via microwave communication requires antennas exceeding 20 m in diameter, which are both costly and technically challenging.

The most advanced alternative is laser communication, which drastically reduces the required aperture size for optical telescopes. On the lunar surface, a 0.5-m telescope can achieve data rates exceeding 1 Gbps. However, the Earth's atmospheric interference—particularly during bad weather or cloudy days—can disrupt laser signals. Currently, there are two solutions: (1) Multi-Station Redundancy. Deploy multiple ground-based laser stations across the Earth. Select stations with clear weather conditions to ensure uninterrupted service. When the number of laser ground stations is large enough, they can ensure that at least one of them is under good weather conditions suitable for communication. (2) Hybrid Relay via Geostationary Satellites. The Moon-satellite link adopts laser communication, while the satellite-ground link adopts conventional high-data-rate microwave to ensure all-time and all-weather connectivity [2].

An alternative approach involves constructing a 20+ km stratospheric observation tower topped with a laser communication terminal. This tower would also function as a stratospheric atmospheric and astronomical observatory, offering high-quality scientific data similar to the orbital observatories by avoiding tropospheric interference. But compared with major space telescopes, the construction cost is prohibitive, which requires joint funding as a national large-scale scientific facility.

8.5 Management of Time Differences

The Moon's day-night cycle lasts approximately 28 Earth days, making it impractical to align human schedules with lunar daylight periods. Instead, future lunar tourism facilities must adopt Earth-based timekeeping that is synchronized with visitors' home countries.

This approach offers several advantages: aligns with tourists' biological clocks, facilitates real-time communication (voice/video) with the ground during compatible waking hours, ensures that the Earth's visible phase matches tourists' local daytime, and distributes facility usage by defining different activity periods, hence reducing the hustle and bustle in limited communal spaces. However, this requires 24/7 staffing and increases labor demands.

8.6 Synergy Between Tourism Facilities and Research Stations

Lunar tourism operates as a commercial space venture, funded through private investment and sustained by operational revenue. Meanwhile, government-led lunar exploration aims to establish permanent research stations. Strategic integration of these efforts yields mutual benefits. Government missions prioritize scientifically significant landing zones over scenic value, but share identical infrastructure needs with space tourism, such as power generation, communications, and life support systems in pressurized habitats.

From the perspective of lunar tourism development, the highest demand is sightseeing and visitor experience. But the high cost of building complete energy, power, communications and life support systems in pressurized modules and the long period to recover the cost necessitate the joint investment with the government.

How do they reinforce and integrate with each other? First, a cost-sharing opportunity lies in infrastructure construction. Tourism facilities require larger-scale, continuous operation, including lunar nights—prime Earth-viewing period. Therefore, research stations could share infrastructure costs and use some hotel rooms to accommodate them. In addition, the dining halls and meeting facilities can also be shared.

Second, collaborative development of lunar landers. The exploration missions have different requirements for landing areas. The government and space tourism companies can work together to modify tourist spaceships to transport scientists to remote sampling sites and return samples to hotel-based labs, eliminating the need for separate polar bases.

Third, unsealed sample repositories or even in situ spectrography labs can be built within the lunar hotel to maintain pristine conditions for analysis. Therefore, it is not necessary to bring most of the samples back to the Earth. Scientists would access these samples via airlocks in standard EVA suits.

Finally, for studies with no site-specific requirements, such as lunar astronomy and Earth observation, the location of tourism facilities could be directly utilized.

References

1. D. Rapp, *Use of Extraterrestrial Resources for Human Space Missions to Moon or Mars*, 2nd edn. (Springer, 2013)
2. L.N. Binh, *Space Photonic Communications*, 1st edn. (CRC Press, 2025)

Chapter 9
Cost and Market Analysis

Space tourism in the real sense should be sustainable: it continuously attracts new tourists, keeps the market dynamic, and fosters ongoing healthy development. Therefore, the number of tourists should not be limited to a few or even several hundreds, but should expand significantly.

When analyzing the market, we should start by estimating and identifying this scale to determine the market's potential size, and then work backward to assess the required infrastructure scale and cost constraints for space tourism.

9.1 The First Wave of Potential Clients

In countries with advanced space programs, such as the USA, Russia, Japan, China, and some European nations, as well as countries that prioritize space science education, there is no shortage of space enthusiasts. These individuals are the most ardent supporters of space tourism, even if many cannot afford to purchase a ticket to space. For them, the opportunity to travel to space even once in their lifetime may be their ultimate aspiration.

When middle- to high-income individuals develop a desire to travel internationally, the leisure expenses in their disposable income typically reach tens of thousands of dollars per year. Subsequently, some among them may seek out extreme travel experiences—such as visiting research stations in Antarctica, setting foot on the Svalbard Archipelago in the Arctic, venturing into wildlife reserves in Kenya, or conquering Mount Everest and other continental peaks. These travelers are willing to spend ten to a hundred times more than average tourists, with the travel expenses reaching hundreds of thousands of dollars. They represent the first wave of potential space tourists.

However, space tourism demands even greater expenditure. Most people might afford it only once in their lifetime rather than traveling to space annually. Thus, we can consider their 10-year budget for adventure travel as the upper limit—setting

J. Wu, *Space Tourism*, SpringerBriefs in Applied Sciences and Technology,
https://doi.org/10.1007/978-981-92-0883-8_9

the price of a single space trip at one million USD. While this remains prohibitively expensive for many, the number of potential tourists across the globe willing to pay this price could easily reach a thousand per year or even more.

Chapters 11 and 12 will delve into the design of space tourism experiences. This chapter focuses on the motivations of the first wave of potential travelers based on fundamental demands.

They are adventurers, risk-takers, and life challengers. The pioneers of space tourism will likely be highly accomplished individuals. The first seven private astronauts to visit the ISS were all successful entrepreneurs. If ticket prices drop from tens of millions to a few million—or even $1 million—the pool of eligible candidates would expand across nearly all professional fields, including well-paid elites in business, fashion, entertainment, and sports. At their career peak, they seek unparalleled challenges, pushing the boundaries of human experience and physiological limits. Space travel could be their ultimate dream.

The first batch of space tourists also includes affluent retirees. Free from financial pressures, with permanent assets and substantial pensions or investments, their unmet desire is the long-held dream of space travel. Traveling with friends, they see it as an integral part of a fulfilling life.

New wealthy individuals in emerging economies, particularly those who accumulated a fortune through land appreciation or real estate, now possess convertible assets worth millions. Selling one of their houses can cash out millions at any given time. For them, space tourism is a unique way to distinguish themselves socially.

China, in particular, harbors a vast reserve of such potential customers. As early pioneers successfully venture into space, market maturity will grow, attracting clients from diverse backgrounds [1].

9.2 Revenue and Constraints

Based on the costs of current government space launch vehicles, sending 1 kg of payload to Low Earth Orbit (LEO) costs approximately $10,000. However, with the rise of commercial spaceflight, this cost has dropped to $5000 or even lower when using commercial launch vehicles.

For space tourism, we assume that the round-trip transportation system, including launch vehicles and spaceships, can be procured commercially without dedicated development by space tourism companies. To send one passenger to LEO, the payload must include not only the tourist but also essential life-support supplies, bringing the minimum mass per passenger to no <100 kg. Thus, the baseline cost to send one person to orbit is around $500,000, excluding expenses for reentry and landing. Of course, the return cost is significantly lower than the cost of sending tourists into space.

This pricing reflects the current low launch cadence and limited mass production. As space tourism thrives, costs will decrease further.

If the ticket price is set at $1 million per passenger, revenue would cover costs and even achieve profitability. However, a mere orbital flight—without docking at a space station like the ISS—would offer only a brief, monotonous experience, limiting its appeal to more potential customers.

Given that the ISS relies on government-funded maintenance at high costs, the most urgent issue for orbital tourism is to build a commercial space station (or "orbital space hotel") as the primary destination. To tap into the market's potential in terms of the number of customers, annual tourist capacity must expand from dozens to hundreds or even thousands.

If there is a six-person spaceship (one pilot + five tourists) for a 6–7-day trip, including 2–3 days aboard the orbital hotel, and only one launch is scheduled per day, the orbital space hotel must accommodate 20–30 people (mostly tourists plus two to three staff). With five tourists going into space per day, the monthly capacity reaches ~150, or ~1800 annually. With each paying $1 million for transport alone, this creates an $18 billion/year market. Therefore, the key constraint is the capacity of the orbital space hotel. Further growth hinges on expanding the hotel's capacity—either by enlarging it or building additional orbital hotels.

9.3 Cost of Orbital Space Hotels

So far, we have estimated only the total revenue and basic round-trip transportation costs, without accounting for the expenses of constructing an orbital space hotel or other space tourism infrastructure. This section examines the cost to build a Low Earth Orbit (LEO) space hotel.

A straightforward calculation method is based on launch mass. An orbital hotel is essentially a space station in LEO, despite the fact that it wouldn't host scientific experiments. It requires the same level of safety, life-support systems, and communication/control capabilities as the ISS. Thus, we can take the ISS's assembly process for reference, including the number of its launches and the launch masses.

The ISS weighs over 400 metric tons and provides nearly 1000 m^3 of interior space. It is assembled through dozens of launches. Assuming 30 m^3 of living space per person, a hotel accommodating 20–30 people would need 600–900 m^3 of internal space. Scaling down the ISS's mass proportionally, the hotel's launch mass would be 250–360 metric tons. At current launch costs ($5000/kg), sending this into orbit would require $12.5–18 billion.

However, inflatable habitat modules currently under development for commercial use could drastically reduce this cost. If deployable structures become viable, expenses would drop significantly.

Beyond the cost of launching, we must consider R&D expenses. Commercial satellite development typically costs $10,000–50,000 per kilogram (e.g., a 1000 kg satellite costs $10–50 million). By extrapolating, we can assume that a 250-ton hotel's R&D would range from $2.5 to $12.5 billion. Given that a space hotel, while massive, is less complex than a satellite, we'll adopt the lower estimate: $2.5 billion for R&D + $12.5 billion for launch = $15 billion in total. This is based on leveraging the existing ISS-derived technologies rather than starting from scratch.

Other costs include ground control, mission operations, TT&C, and communications, which typically account for 10–20% of total mission costs.

Assuming a 15-year operational lifespan and a 5-year cost-recovery period, the $15 billion investment would require $3 billion in revenue per year. Spread across 1800 annual tourists, the cost distribution is $0.44 million per person. If tourists stay in the orbital hotel for 2–3 days, the additional cost is $0.5 million. Thus, a 6–7-day orbital trip would cost roughly $1.5 million per passenger.

As previously analyzed, this pricing aligns with the first wave of affluent adventurers. Once demand scales up, costs will drop further, making space tourism sustainable and profitable.

9.4 Cost Differences Between LEO Tourism and Lunar Tourism

As the analysis in Chap. 11 and Chap. 12 shows, lunar tourism holds far greater appeal than Low Earth Orbit (LEO) space tourism. However, expanding infrastructure from an orbital space hotel to a lunar hotel introduces significant additional costs.

The first step is developing a dedicated Earth-Moon Shuttle Spaceship (EMSS) to transfer tourists from the LEO space hotel to the lunar orbit. Unlike crew capsules launched from the Earth, this spaceship would operate only between the orbital space hotel and the Moon, enabling repeated reuse. The development cost of this shuttle spaceship should be no less than a few hundred million US dollars, even in commercial mode. Furthermore, we should add the one-time launch cost to LEO: ~$100 million.

Its operational lifespan is more than 5 years. To mitigate risks, the vehicle's capacity would be in line with the spaceship operating between the Earth launch pad and the LEO: one pilot + five passengers. For cost amortization, the annualized development and launch cost is around $100 million/year. If every round trip lasts for 6 days, the annual number of round trips would be around 60. Therefore, the cost per trip is $1.7 million and the cost of a passenger is $340,000. Note that we need to add the 30% operational costs (fuel, maintenance). In case the tourists in LEO space hotel want to future venture out to the Moon, they need to pay additionally ~$500,000 per passenger, just 1/3 of the total LEO trip cost, not the 3 times pricing some companies currently quote.

The second step is building a lunar orbital station for the purpose of refueling the Earth-Moon shuttle spaceship, as well as providing a platform to appreciate the Moon and look back at the Earth. However, the function of this space station is for tourists' transfers and short stays, rather than for accommodation like space hotels in the Earth's orbit. Therefore, the scale of this lunar orbital space station can be smaller, and the corresponding costs will also be lower.

The most expensive component is the permanent lunar hotel, requiring continuous power systems for the 14-day lunar night. The exact costs remain speculative, but landing on the surface of the Moon and staying at the lunar hotel would multiply trip expenses as compared with orbital tourism. It also demands a longer development cycle and longer periods of construction.

Currently, the pricing of government-facilitated LEO tourism is \$55 million/ticket and the pricing of a commercial lunar orbital tourist mission is \$175 million per flight for multiple passengers. If divided by 12 passengers, the pricing is ~\$14 million/person. Based on the above analysis, the future commercial tourism, with the pricing of SpaceX as a reference, will drive the ticket price down to \$3–5 million/person for orbital trips, which is far more attractive than the current \$55 million/ticket pricing of government-facilitated LEO tourism.

Like LEO space hotels, the construction of lunar hotels will evolve iteratively, with capacity determining market ceilings. Even at inception, the lunar tourism market will span tens to hundreds of billions of dollars.

9.5 Venture Capital and Profits

In the preceding analysis, we have not yet accounted for the role of venture capital (VC) in funding early-stage space tourism, nor the value appreciation generated as both the tourism companies and the broader industry mature.

Venture capital is a primary funding mechanism for startups in emerging sectors—particularly in internet and commercial space industries. Key space tourism segments attracting VC investment include: launch vehicles, crewed spaceship and reentry capsules, LEO space stations/hotels, high-speed data relay communications, Earth-Moon Shuttle Spaceship, lunar orbital stations, spaceship operating between the lunar surface and the lunar orbital station (crewed/uncrewed), lunar regolith fuel extraction, in situ lunar resource utilization, lunar hotel construction, space tourism operations, etc.

VC firms, with deep capital reserves, are typically staffed by experts in technology and finance who are capable of assessing industry trends and entrepreneurial competence. They take high risks by investing in startups. Normally, they acquire minority equity stakes and profit from enterprise value growth—without direct operational involvement.

As a frontier high-tech sector aligned with humanity's future, space tourism inherently promises long-term value increase. However, this hinges on company performance and market adoption.

Space tourism is an open market, with the governments' role limited to policy-making, safety oversight, and legal regulation. Devoted entrepreneurs of the industry compete in a fair play, where market forces ultimately favor those with cost-efficient pricing, cutting-edge technology, and superior service delivery.

Successful companies leverage early-stage venture capital (VC) and later-stage initial public offerings (IPOs) to raise the necessary funding. Due to the appreciation of the company, the investment, as shares, will also appreciate, thus not only recovering the costs but also earning substantial returns. It is evident that there is no need to wait for 5, 8, or even 10 years until profits fully cover all investments to achieve financial balance. Instead, as the company's value increases, later investments continuously make sure that the early investors can exit with robust returns. The space tourism company repays early investors through its increasing value and continuously receives rolling capital support, enabling its sustainable development.

Future space tourism pricing will reflect infrastructure capacity ceilings and market demands. Rather than being constrained by static cost-profit equilibrium models, prices will decline progressively, edging toward mass affordability [2].

References

1. AI Lab for Book-Lovers, *Space Tourism: For All or a Few?* (Xynapse Traces, 2025)
2. E. Cohen, S. Spector (eds.), *Space Tourism: The Elusive Dream*, 1st edn. (Emerald Publishing, 2019)

Chapter 10
Risk Analysis and Mitigation

In 2006, while accompanying a delegation from the China National Space Administration (CNSA) to NASA Headquarters, I observed that a long corridor leading to the administrator's office was lined with portraits of the 14 astronauts killed in the Space Shuttle disasters of Challenger and Columbia. Each day as NASA's top leadership enters their offices, they meet the gaze of these astronauts—a constant reminder of how crucial safety is in human spaceflight.

From the perspective of government space programs, any failure constitutes a "crime" against taxpayers and represents negligence in managing public funds. This cannot be forgiven. Therefore, not just in China but globally, government space missions maintain an uncompromising emphasis on flawless execution—a requirement that persists throughout all mission phases, particularly for human spaceflight. This remains one of the primary reasons for the high costs of government space programs.

However, for commercial space missions, including those involving human safety like space tourism, the considerations and tolerance regarding safety differ fundamentally. Commercial space entities aren't accountable to the general public but specifically to paying customers. When clients voluntarily assume certain risks, industry practitioners, including designers and operators, must take this factor into their approach—optimizing cost-efficiency while implementing reasonable risk mitigation strategies, rather than doggedly pursuing perfection at any cost.

10.1 Civil Aviation vs. Space Tourism

When examining the risks of commercial space tourism, it is inspirational to compare with the early development phases of civil aviation.

The world's first commercial airline, Braniff International Airways, was established in the USA on January 1, 1914. The oldest continuously operating airline under its original name is KLM Royal Dutch Airlines, founded in 1919 with over a century's service. Today, hundreds of aircraft take off and land globally every

J. Wu, *Space Tourism*, SpringerBriefs in Applied Sciences and Technology,
https://doi.org/10.1007/978-981-92-0883-8_10

minute, with over 10,000 planes in the air simultaneously. However, the frequency of fatal accidents is less than one per year. Back a century ago, however, accident rates were significantly higher, claiming even famous figures we all recognize. Why did commercial aviation persist and thrive? [1].

The answer lies in passenger demands—travelers knowingly accept risks when purchasing tickets. This reflects the free market principles. Even today, aviation technology cannot guarantee absolute safety, yet people voluntarily assume risks associated with this mode of travel, thereby sustaining technological and market development. Aboard any major airliner, we may find that a significant proportion (at least 30%, if not 50%) of passengers are tourists, with the remainder traveling either for business or visiting friends/family who are essentially tourists in another form.

The emerging space tourism market mirrors this evolution. A distinct group exists that yearns to leave the planet Earth to experience space travel, even lunar trips, and to look back at our home planet from that vantage point. These pioneers are willing to pay premium prices and accept minute-but-real risks: launch failures, micrometeoroid penetration causing cabin air leak, or life-threatening high-energy particles from space weather events. But they remain undeterred in their cosmic aspirations.

To serve this demand, space tourism operators are engaged to reduce costs while maintaining safety thresholds. They facilitate these dreams with full transparency about residual risks. When asked "We've done everything possible, but risks remain. Do you still want to go?", the normal response is: "Yes, we understand and accept the risks. Let's go ahead."

Obviously, the overwhelming principle here is not perfection, but the optimal balance of safety, cost, efficiency, and customer satisfaction. Space tourism practitioners equally share risk accountability—much like commercial airline pilots who cannot eject but must remain with their passengers through any outcome.

10.2 Risks Beyond Technologies

In addition to technical risks, operators must consider health-related risks that tourists may encounter during space travel. Therefore, space tourism companies must conduct comprehensive physical and psychological evaluations of passengers to prevent emergencies during flight.

Physiological risks primarily involve assessing passengers' adaptability to hypergravity during launch and microgravity effects in orbit. Under hypergravity, blood circulation faces resistance against gravitational forces, which is why astronauts recline during launch to prevent blood pooling. In microgravity, blood redistributes too easily toward the head, potentially causing congestion. Thus, pre-flight physiological screening focuses on cardiovascular parameters.

Psychological risks center on passengers' ability to cooperate in confined spaces. Space stations or orbital hotels have limited space, requiring all tourists to maintain a collaborative attitude. One disruptive individual could significantly impact others' experiences in such tight quarters [2].

Focused ground training can help screen potentially problematic passengers. However, the Gagarin Cosmonaut Training Center's 6-month program for private ISS visitors (2001–2009) is too lengthy for tourists. Since space tourism isn't astronaut training, new physiological and psychological standards should enable efficient evaluation, screening, and essential training within minimal timeframes.

Additionally, orbital hotels must implement medical treatment and emergency protocols, with passengers pre-authorizing necessary interventions under emergency situations.

10.3 Risk Mitigation

Commercial aviation mitigates crash risks primarily through insurance, a model applicable to space tourism. Like extreme activities such as bungee jumping and skydiving, participants must sign waivers and purchase appropriate insurance.

Normally, robotic space mission insurance premiums reflect launch success rates in the past 10 years. Ninety-nine percent success (one failure/100 launches) means that the insurance premium is ~1% of the mission cost. Ninety percent success (one failure/ten launches) means that the insurance premium is ~10% of the mission cost.

Space tourism can adopt this model, applying comparable percentages to passenger life insurance pools to identify the insurance premium for the launch campaign. Premiums for much lower-probability risks, such as space debris and solar storms, decrease significantly given their predictability.

Ethically, space tourism operators must require baseline insurance included in ticket prices. Additional premiums on top of this can be chosen by the tourists at their own discretion.

References

1. A.P. Dobson, *A History of International Civil Aviation: From Its Origins Through Transformative Evolution*, 1st edn. (Routledge, 2017)
2. L.B. Landon, K.J. Slack, E. Salas (eds.), *Psychology and Human Performance in Space Programs: Research at the Frontier* (CRC Press, 2024)

Chapter 11
Immersive Experience and Service Design for LEO Space Tourism

Space tourism is a unique experience that, like other forms of tourism, requires thoughtful service design. Well-designed content can make the entire journey incredibly engaging. It is attractive not only to those who have never been there, but also to those who have been there. Knowledge-intensive designs allow tourists to gain unexpected insights and learn things inaccessible to those who haven't ventured into space. Profound design leads to amazing space encounters that can purify the soul and transform perspectives. This chapter will explore how space tourism organizers should design itineraries, activities, and educational content to ensure that participants not only gain knowledge but also experience spiritual purification and personal growth.

11.1 Ground-Based Training and Selection of Ourists

Space tourism begins with ground-based training. Although tourists have no work obligations in space, they must adapt to environments that are vastly different from the Earth. If passengers discover in the mid-flight that they cannot physiologically or psychologically tolerate these conditions, it could endanger them as well as others who share the trip. Thus, ground training serves two purposes, i.e., enhancing passengers' adaptability to the new environment, and identifying individuals unlikely to tolerate space conditions so as to advise them to withdraw before departure. Otherwise, once launched, it is too late.

Some may argue that everyone can endure space environments. If not, how will humanity leave the cradle of the planet Earth? This is a crucial question. Over nearly two million years of evolution, human beings have adapted to the land environment, and can stay for a long time in the ocean, and are now venturing into space. This is a slow environmental selection process. Those not suitable to such changes were gradually phased out. Leaving the cradle of the planet Earth won't happen in decades. Instead, it may take a century or more. Through repeated selection

J. Wu, *Space Tourism*, SpringerBriefs in Applied Sciences and Technology,
https://doi.org/10.1007/978-981-92-0883-8_11

processes, those who can adapt to gravity shifts and microgravity will emerge as pioneers.

Besides teaching essential space knowledge, ground training focuses on overcoming physical and psychological challenges after leaving the planet Earth. A key physiological hurdle is space motion sickness, which differs significantly from car sickness or seasickness. It is worth noting that people who frequently experience car sickness or seasickness may not suffer from space motion sickness; conversely, those who do not get car sickness or seasickness may still experience space motion sickness. Even astronauts, who are normally former pilots with strong vestibular systems, can suffer from space motion sickness. Motion sickness stems from inner ear fluid reacting to unexpected body movements. But space motion sickness arises from abnormal responses of otoliths (tiny calcium crystals in the inner ear). Constant head movements bend ~20,000 hairlike neurons on otoliths. On Earth, gravity bends these neurons, helping the brain discern orientation—like a hypersensitive accelerometer. In microgravity, the absence of gravity on otoliths tricks the brain into sensing a fall, while the eyes confirm stability. This conflict causes disorientation, even nausea or violent vomiting [1].

Symptoms typically subside and disappear within 1–2 days as the brain ignores otolith signals and relies on visual cues. By that time, the passengers can have better control of their bodies and better orientation. Experienced astronauts adapt quickly by recognizing microgravity and dismissing misleading vestibular input. Ground-based training cultivates this adaptability. Those who fail to adjust may be advised to quit space travel.

Additional physiological screening should identify and exclude individuals with hypertension or heart disease. Psychological screening mirrors screenings for high-stakes, teamwork-dependent professions.

11.2 Launching

As the tourists, clad in spacesuits, settle into the seats of the launch-ready spaceship and fasten the safety harnesses, a mix of exhilaration and nervousness is inevitable. This presents a prime opportunity for launch safety education.

While emergency abort procedures for launch failures must have been covered during the ground-based training, reiterating these procedures at this moment ensures better understanding. Therefore, during this waiting period, every detail of the abort sequence should be reviewed again. This safety briefing also serves to calm nerves and boost the confidence of the passengers.

Passengers should understand that all contingencies have been accounted for, and their safety is fully guaranteed even in emergencies during the launching. Then comes the countdown, ignition, and liftoff.

The rocket's tremendous power, accompanied by violent vibrations, thunderous noise, and crushing g-forces, makes this the most thrilling phase of space tourism. Having previously witnessed launches as spectators, they now personally experience hundreds of tons of rocket gradually leaving the launch pad and accelerating. They can directly sense the vehicle's energy as they depart the Earth's surface and reach orbital velocity.

Though pressed deep into their seats with blood nearly stagnant from acceleration, sheer excitement dominates their feelings. This is the moment to let passengers fully immerse themselves in the excitement of high-g acceleration and the rollercoaster-like weightlessness during stage separation. It will be an unforgettable experience. Any broadcasting or announcement is unnecessary at this moment.

Key educational components during this phase include: Earth's gravity, universal gravitation, velocity versus acceleration, and the first cosmic velocity.

11.3 In Orbit

When the last rocket stage shuts down, the spaceship enters sustained microgravity, marking successful insertion into the Earth orbit. While ground training includes brief microgravity simulations, even parabolic aircraft flights for some tourists, none of these simulations can compare to the continuous microgravity of actual spaceflight.

Initially, beyond vestibular disorientation, passengers experience the movement of bodily fluids to the head as the heart continues pumping blood as if under the Earth's gravity. In this situation, it is no longer necessary to fight 1 m of pressure to return blood to the heart from extremities, such as the feet and legs.

Cardiac output gradually decreases during the passengers' adaptation. Like space adaptation syndrome, these effects typically resolve within 1–2 days, justifying minimum 2–3-day stays for proper acclimatization to prepare for a wonderful experience in space.

Daily activities become novel experiments: eating, sleeping, and going to the toilet all require microgravity adaptations. Food comes in sealed packages to prevent floating debris. Drinks are consumed through straws rather than open containers. Sleepers secure themselves to walls with safety belts to avoid floating in sound sleep, eliminating traditional "lying down" orientation.

While astronauts have to undergo complex hygiene routines, tourists may optionally experience space showers during short stays. Passengers need to take a shower in a separate compartment, not only for privacy but also to prevent water droplets from spreading to other areas of the space station. The vacuum systems will remove all residual droplets after towel drying.

Body waste management proves most challenging. To facilitate more convenient toilet use in space stations, various extremely expensive space toilets have been developed. Their primary working principle involves using negative pressure to quickly suck waste into pipes, which either enters the recycling system or will be harmlessly processed before being released into space.

Beyond daily necessities, space tourism designers must design various entertainment activities, such as floating water globules, midair acrobatics, object toss games, and corridor "swimming." It is certain that with the flourishing of space tourism, some unique activities will keep emerging, such as microgravity ping-pong or hoop maneuvers.

Educational components include: mass vs. weight, centripetal force and centrifugal force, fluid dynamics, and physiological adaptations to microgravity, including blood, muscle, and bone.

11.4 Space Hotels

A 2–3-day stay on a fully interconnected orbital station feels less like ideal space tourism and more like experiencing an astronaut's working environment during national missions. Space tourism in the real sense means having private rooms in a space hotel with Earth-viewing windows.

Currently, several companies are developing microgravity "couple's suits"—elastic garments that connect two people, creating restorative tension when they drift apart in weightlessness.

11.5 Viewing the Earth

Orbital hotels will feature large observation windows, possibly in every room, for viewing the Earth's rapidly passing landscapes below. During each 90-min orbit of the space hotel, passengers witness a complete day-night cycle, and can see the Sun rise and set behind the Earth's curvature.

Of course, to watch the Sun directly with the naked eye is impossible. In space, due to the absence of atmospheric filtering, sunlight becomes brighter and more glaring. The atmospheric limb appears breathtaking with a delicate blue halo encircling the curved Earth, especially during orbital sunrise/sunset when the atmosphere is illuminated by sunlight from behind. If the passengers are lucky, they might also have the chance to see the beautiful aurora.

Thanks to the development of the internet, ground-based viewers can now watch videos filmed on the space station, including 24-hour live broadcasts. However, this still differs greatly from personally experiencing the beauty of the Earth from space. Especially when tourists fly over their homeland or the city they live in, their emotions must be very special. They can see the country they love where their loved ones live. Therefore, this must be included in the itinerary and arranged as a highlight of the entire trip.

The passengers can see the night side illuminated by city lights, and megacities glowing brightly with suburban sprawl. Thunderstorms, ever-present somewhere on the Earth, display spectacular lightning flashes that illuminate entire cloud systems, making them easily recognized. Before tourists had time to appreciate them carefully, the spaceship was already back on the day side of the Earth.

Educational components include: atmospheric science, ionosphere, geomagnetism, lightning, plate tectonics and oceans, and other geophysical knowledge.

11.6 Communication with Loved Ones

Space communication with family and friends is another activity that must be designed into the agenda. Even as space tourism matures, venturing into space at the cost of a million-dollar ticket may remain a once-in-a-lifetime experience. Real-time sharing of this experience—ideally via video calls—carries tremendous emotional value.

Due to the limited communication capabilities of the entire space hotel, additional calls, especially video calls, beyond the scheduled free call timeframe should be charged. This is because space hotels orbiting the Earth at high speeds cannot maintain direct contact with the ground at all times, and communications require satellite relays. Space tourism companies can lease existing satellite networks rather than building dedicated systems at their own cost.

Education components include: RF/laser communications, data relay systems, data rates, and signal modulation/demodulation via electromagnetic waves.

11.7 Extravehicular Activities

Incorporating spacewalks into LEO space tourism presents significant challenges and inherent risks.

The EVA procedure is complicated [2]. First, the passenger should enter an airlock chamber to wear an EMU (Extravehicular Mobility Unit). After gradual depressurization and complete vacuum is achieved, the chamber will open and the passenger is ready for extravehicular activities. Then, the passenger returns to the airlock chamber, closes the hatch, pressurizes the interior of the chamber to one atmosphere, removes the EMU, and returns to the interior of the space station. This labor-intensive process demands dedicated crew supervision, resulting in premium pricing that should be covered by passengers who want to give it a try.

Outside the spaceship, the assigned EVA guide, a one-on-one space tour guide, must constantly keep an eye on the spacewalking tourist to ensure safety. In orbit, any push against the station propels the tourist outward, heading into deep space. Should this occur, the tourist would have no means of returning, as there are no footholds in the emptiness of space to push against. This is unlike the situation in a swimming pool, where one could paddle back. The result would be permanent orbital drift until the limited oxygen and energy reserves in the spacesuit are depleted, leading to a tragic end and transforming the individual tourist into space debris.

For this reason, every spacewalk must involve a safety tether, secured via a carabiner to handrails or rings on the station's exterior. This tether also ensures that the tourist's electrical potential matches that of the station's surface, preventing electrostatic discharge. The accompanying guide's primary responsibility is to prevent the tourist from drifting away and becoming space debris.

Of course, there are self-propelled spacesuits equipped with thrusters that could allow a tourist to maneuver back to the station. However, for non-professional tourists, operating such a powered suit is accident-prone. Therefore, extravehicular activities should only be conducted using non-propelled suits.

Spacewalks must only be scheduled after tourists have fully adapted to microgravity, never while they are still experiencing space sickness. Vomiting inside a spacesuit would have severe and hazardous consequences.

During an EVA, educational components include the operating principles of an airlock, the design and functionality of extravehicular suits, and the space environment outside the space station.

11.8 Return to the Earth

Returning from an LEO space station is an equally thrilling experience. After days in space, tourists are ready to reenter the atmosphere to embrace Mother Earth. They enter the shuttle spaceship operating between the Earth and the orbital space station, which then undocks from the station and awaits the optimal reentry window. The retro-rockets fire to decelerate, plunging the spacecraft into the plasma-generated "blackout phase," where intense friction with the dense atmosphere causes violent shaking. Through the windows, the passengers can see a fiery red glow engulfing the view. During the blackout, the communications are temporarily lost for several minutes.

First, the drogue parachute of the return capsule deploys, followed by the main chute. Once the main chute is fully in swing, almost all turbulence ceases, and the descent capsule settles as if cradled by Mother Earth. The return to gravity reminds tourists of Earth's persistent gravitational pull. They are now readjusting themselves from microgravity. No announcements are necessary during this phase. The immersive experience itself is the highlight of the journey.

Upon touchdown, passengers feel their full weight press heavily into their seats, making even slight movements difficult. A 30-min to 1-h readjustment period is required for the cardiovascular system to readapt, restoring the ability to pump blood from the legs back to the heart and brain. Gradually, under crew guidance, tourists may move and rise from their seats.

Outside, rescue teams and loved ones gather as the capsule hatch opens. Each passenger is carefully assisted out of the capsule and enters a transport vehicle, marking the successful conclusion of their space voyage.

Key educational components during reentry include: atmospheric layers, the blackout effect, hypersonic heating, parachute mechanics, meteorology, primary and secondary landing sites, and gravity re-adaptation.

References

1. J.R. Lackner, Motion sickness: more than nausea and vomiting. Exp. Brain Res. **232**(8), 2493–2510 (2014). https://doi.org/10.1007/s00221-014-4008-8
2. D.J. Shayler, *Walking in Space: Development of Space Walking Techniques (Springer-Praxis Books in Astronomy and Space Sciences)* (Springer, 2004)

Chapter 12
Immersive Experience and Service Design for Lunar Tourism and Beyond

Space tourism is far more than just experiencing weightlessness. In fact, microgravity is merely an inevitable condition of traveling to space. True space tourism involves visiting another celestial body. As our nearest celestial neighbor, the Moon serves as the primary focus of this chapter, followed by prospects for visiting other planets and their moons in the solar system.

12.1 Earth-Moon Travel

After experiencing LEO tourism, the Moon naturally becomes the next aspirational destination. The journey spans 380,000 km, and with current spacecraft propulsion capabilities, the optimal flight time is 2–3 days. Thus, lunar tourism must incorporate meaningful activities during this transit period rather than leaving passengers passively waiting.

As the voyage begins, the receding Earth offers boundless wonder and inspires profound contemplation. Passenger seats are arranged in a dispersed layout, so the capsule windows face different directions. To optimize the viewing experience, the spacecraft should slowly rotate along its axis, ensuring every traveler gets a chance to periodically access every viewing angle. Through this rotation, each passenger witnesses our home planet Earth gradually shrinking into the distance. At this stage, still relatively close to home, the sensation of departure may not yet be profound.

Since continuous observation isn't possible as the view alternates due to the spacecraft's rotation, the Earth's own spin may go unnoticed. Yet, if the visible hemisphere happens to be their home country, whether in daylight or darkness, the passengers could tell the familiar contours of their cities. However, when the entire Earth comes into view—no longer a partial disk but a complete planet—the sight is awe-inspiring.

By now, the spaceship has moved beyond the Earth's shadow into interplanetary space, with sunlight perpetually illuminating one side of the capsule. The slow

J. Wu, *Space Tourism*, SpringerBriefs in Applied Sciences and Technology,
https://doi.org/10.1007/978-981-92-0883-8_12

rotation also ensures even thermal distribution across the spaceship. When a passenger's window faces the Sun, the glass must automatically filter intense solar radiation to protect their eyes. Providing sunglasses further enhances the opportunity to observe the Sun and learn about solar activities.

When the window faces deep space, passengers behold a star-studded cosmos, with the Milky Way shining more vividly than any observation from the Earth. Depending on the orbital position, the onboard audio system can guide passengers to identify Mercury, Venus, Mars, Jupiter, and Saturn, and can share other astronomical knowledge.

The true highlights, however, are gazing back at the planet Earth and watching the Moon. Over the journey, the Earth's sunlit portion may appear as a crescent or nearly full, depending on the lunar calendar date and the Sun-Earth-Moon alignment. Similarly, the Moon's illumination varies. Travelers departing around the 15th day of the lunar month will witness a full Moon but see Earth as a slender crescent (or Earth crescent), since the spacecraft flies between the two bodies, with the Sun lighting up the Earth's nightside from behind.

It is recommended to postpone lunar tourism until breakthroughs in in situ fuel extraction on the Moon are achieved. At that stage, Earth-Moon Shuttle Spaceship could maintain a gentle acceleration of 0.1–0.16 g, with seating orientation aligned with gravity, so that the thrust direction mimics "downward" gravity. Though weak, this acceleration would allow passengers to eat and drink at tables almost as easily as on the Earth, thus vastly improving comfort compared to weightlessness.

Key educational components during Earth-Moon transit include Earth-Moon orbital mechanics, solar observations, five planets and the Milky Way.

12.2 Orbiting the Moon

During the latter phase of the journey, the Earth-Moon Shuttle Spaceship will reorient itself to decelerate, eventually being captured by the Moon's gravitational field to enter the lunar orbit. At this point, passengers can observe the Moon at close range. Through the windows, they will see gray impact craters of varying sizes and the darker, smoother lunar Mares, which are so named despite being completely waterless. The lunar Mares are composed of fine regolith with low albedo, which explains their somber hue. This is a celestial body whose surface morphology has remained virtually unchanged for billions of years. Though the Moon is the natural satellite of the planet Earth and part of our planetary system, its topography and physical properties differ dramatically from our home planet.

When the spacecraft passes over the sunlit side, the lunar surface appears in stark contrast; over the night side, it plunges into near-total darkness. However, if the Moon is positioned between the Earth and the Sun, sunlight reflected from the Earth will faintly illuminate the lunar terrain.

No passenger should miss the pivotal moment when, as the spaceship is orbiting the Moon, the blue Earth appears to rise gradually above the lunar horizon—a phenomenon dubbed *Earthrise* (as compared with sunrise). Why do humans cherish sunrises on the Earth? Because the Sun brings energy, vitality, and the promise of a new day. But during Earthrise, passengers will confront a profound realization that the rising celestial body is home. The planet Earth embodies life, civilization, wisdom, and peace. Witnessing Earthrise is the emotional apex of lunar tourism. Similarly, there exists *Earthset*, when our planet dips below the Moon's horizon.

When frequent lunar landings have been materialized, the lunar orbital tourism will mainly take place in a lunar orbital station. The Earth-Moon shuttle spaceship docks with the lunar orbital station to transfer passengers: delivering those bound for the lunar surface and collecting the returning travelers from the lunar spaceship operating between the lunar surface and the lunar orbit.

Key educational components at this stage include theories of lunar origin, geological evolution of lunar features, physical parameters of the Moon.

12.3 Lunar Surface Tourism

Within 10–20 years of sustained space tourism development, lunar surface tourism is set to become a reality in the near future. It is therefore imperative to conceptualize the lunar tourist experience now. Moving beyond Apollo-era challenges, we focus directly on tourists' experiences within pressurized lunar habitats.

First, let's imagine how to walk in 1/6 of Earth's gravity. Within the pressurized lunar hotel, free from the constraints of spacesuits, tourists can fully experience lunar walking. While this may sound simple, it's actually quite complex. Although lunar gravity is only 1/6 of Earth's, a tourist's body mass remains unchanged. This means that when tourists are moving vertically, like jumping, they feel dramatically different, and when they are moving horizontally, like turning around, they require the same inertial force as on Earth. For the tourists who are familiar with the gravitational force on the Earth, this is quite unusual and requires time for the body to adapt to the Moon's gravity. This discrepancy often leads to falls (though injuries are unlikely due to reduced gravity) as tourists adjust to combined movements like jump-turns.

Second, the design and construction of lunar hotels must prioritize Earth-viewing opportunities. An underground bunker-style hotel without windows would significantly diminish the experience. In that case, tourists can only watch the Earth when they walk outside the hotel wearing extravehicular spacesuits, so the purpose of lunar tourism would be greatly compromised. Optimal designs should include Earth-facing windows in every room, giving the tourists opportunities to gaze and appreciate the Earth. As previously mentioned, the location of the hotel needs to be well-designed, with an appropriate angle for observing the Earth.

Last, it is important to establish proper Earth-Moon Communication. Real-time audio links with the Earth, if not video links, synchronized with Earth-gazing, provide profound emotional connections. It should be noted that due to the long Earth-Moon distance, 1.3-s signal delay each way creates almost 3-s conversation gaps.

The talk with loved ones on that distant blue marble, some 380,000 km away, proves profoundly transformative for the tourists. These basic communication packages should be included in the ticket price (with a ceiling on the time duration of talks), while premium options should also be available for tourists who want to talk more.

Extravehicular activities, while highly desirable, cost additional energy and resources [1]. They should be available as premium options. Guided Moonwalks and rover tours recreate the Apollo moment for the tourists, when the Apollo astronauts first landed on the Moon in 1969–1972. Underneath the feet of the tourists lies the lunar Mare that has remained undisturbed for hundreds of millions of years. Such a sense of time and history was described by Buzz Aldrin, the astronaut who landed on the Moon with Neil Armstrong aboard Apollo 11: "desolate for the eons of lifelessness".[1]

The defining moment comes when a tourist stands alone facing primordial lunar terrain with the lunar rover and other passengers behind. As Aldrin described, you would contemplate humanity's magnificent achievement and question how you are being here as part of this achievement. When the Earth hangs before you in the lunar sky, the emotional impact differs profoundly from viewing it through a window. The stark contrast between the Earth's vibrancy and the Moon's desolation heightens one's sense of stewardship toward our fragile planet and awakens our awareness to protect our mother planet.

Lunar tourism operations are constrained by the Moon's day-night cycle and corresponding energy limitations, which also determine the Earth's visibility. During the lunar daytime, the Sun and the Earth align in the sky. At lunar noon, the Earth becomes nearly invisible, showing only faintly its night side against the solar glare. The exception is during lunar solar eclipses when the Earth blocks the Sun. It is a great time to observe the Earth, particularly for studying atmospheric phenomena.

Lunar night presents optimal Earth viewing conditions. As lunar night progresses toward midnight, more of the Earth's sunlit side becomes visible. At lunar midnight, coinciding with the Earth's full Moon phase, "Full Earth" illumination occurs. This represents the premier Earth-gazing opportunity.

Here, the Earth appears 3.67 times larger than the Moon (as seen from the Earth), as the Earth's diameter is 3.67 times larger than that of the Moon. The Moon's surface, even without direct sunlight, would also be brightly illuminated by the sunlight reflected from the Earth.

Educational components include low-gravity, mass vs. weight, celestial mechanics of the Earth-Moon-Sun system. It is worth noting that tourists can fully leverage the opportunity to understand the significance of the Earth as our mother planet and gain profound philosophical revelations to shoulder our responsibility to protect our home planet.

[1] Buzz Aldrin Interview. [1998-11-17] https://teacher.scholastic.com/space/apollo11/interview.htm

12.4 Future Space Tourism in the Solar System

Beyond the Moon, our solar system offers extraordinary destinations for future space tourism. As its only intelligent species, humanity bears both the privilege of exploration and the responsibility to protect it.

When people think of human expansion beyond the Earth, Mars is likely to be the first destination that comes to mind, and there is growing momentum to immigrate to Mars [2]. However, I believe it is more appropriate to frame this endeavor as tourism to Mars rather than immigration to Mars. Of course, tourism would require the construction of essential infrastructure, such as ISRU (In-Situ Resource Utilization) systems for fuel production and bioregenerative life-support habitats.

One might argue that these developments constitute the foundational infrastructure for immigration to Mars. However, immigration itself should not be the ultimate purpose of building these facilities. The first occupants of these facilities—whether for short-term or extended stays—would primarily be personnel responsible for construction, operation, and maintenance of the infrastructure. Their presence would serve functional purposes rather than permanent settlement. The initial facilities on Mars will primarily support rotating crews of construction and maintenance specialists, who work there temporarily or for longer terms. These professionals would work on rotational schedules between Mars and the Earth.

True "settlers" would require full self-sufficiency. At least for several decades, the Martian economy will center on supporting tourists from the Earth. Therefore, these pioneers on Mars are part of the space tourism industry rather than genuine settlers.

Why Mars? Martian tourism fulfills three innate human drives: scientific curiosity, kinship to the solar system, and stewardship awakening. The Red Planet presents both eerily familiar landscapes and utterly alien natural scenery. Daytime temperatures in deep valleys can rise above the freezing point.

Although tourists still need to carry oxygen tanks on their backs and wear helmets and masks, they would wear streamlined Mars suits, which are less bulky than orbital EVA suits. On Mars, towering mountains far dwarf Mount Everest. The Mars terrain reveals, in front of tourists, the epic geological dramas frozen in time several billion years ago. The desolate vistas serve as cautionary reminders of the Earth's potential future if we don't cherish our home planet.

Beyond Mars, Jupiter's moons offer breathtaking celestial shows. On these moons, tourists can appreciate Jupiter's rising/setting, its Great Red Spot, and swirling atmospheric movement. Io's hyperactive volcanoes produce spectacular lava fountains. Europa's ice-shell resonates with sounds from its deep ocean, and tourists can even go skating on the icy surface. Further beyond, tourists can traverse the belt of small rocky bodies that make up Saturn's massive rings, with those small objects passing by the spaceship one after another. Tourists can also land on the surface of Saturn's moons, admire the rise and set of the enormous Saturn and its rings, and even go boating on the liquid methane seas of Titan.

As technologies related to ISRU and propulsion mature, the spaceship heading toward deep space in our solar system must be capable of continuous acceleration to ensure that passengers on long-distance flights live in a certain gravity environment rather than a microgravity environment. As space life sciences and technology advance, space biotech will provide diversified dining options throughout extended journeys. If the flight duration is still too long, such as requiring several months or even years, future technology could enable humans to enter biological hibernation to endure the lengthy and arduous space travel [3].

If the aforementioned technologies can all be realized, are we still far from traveling throughout the entire solar system and enjoying its scenic beauty?

References

1. G. Heiken, E.M. Jones, *On the Moon: The Apollo Journals* (Springer; Praxis Publishing, 2007)
2. L. David, *Mars: Our Future on the Red Planet* (National Geographic, 2016)
3. C. Wanjek, *Spacefarers: How Humans Will Settle the Moon, Mars, and Beyond* (Blackstone Publishing, 2020)

Chapter 13
Current Imperatives

Since the first space tourist visited the International Space Station (ISS) in 2001, more than 20 years have passed. Yet, space tourism is still primarily focusing on laying foundations through attracting investments and fostering more private space enterprises that leverage government technologies to rapidly enhance capabilities, and establishing New Space R&D systems distinct from traditional government space programs with an emphasis on creativity, failure tolerance, cost reduction, and accelerating technological mastery through iterative testing to learn from trials and errors. The technology breakthrough aims at reducing the costs to enter space, particularly through reusable engine development. In addition, extensive public outreach and the flourishing science fiction inspire younger generations, disseminate scientific knowledge, and cultivate future space enthusiasts.

These efforts have yielded remarkable results, significantly boosting public confidence in space tourism. This chapter examines the current imperatives in space tourism.

13.1 Further Reducing Transportation Cost To and From Earth Orbit

As mentioned in previous chapters, the current launch costs remain at \$3000–\$4000 per kilogram. For a 70 kg tourist who requires some 30 kg life support supplies, this translates to a ticket price of \$300,000–\$400,000, excluding the reentry cost. Apparently, it is imperative to dramatically reduce costs.

Key cost-reduction strategies include full reusability of all launch vehicle stages, ground-based acceleration systems (e.g., electromagnetic launch combined with chemical propulsion) to minimize reusable component risks, parachute recovery of upper stage engines, increased thrust-to-weight ratios to improve payload efficiency.

J. Wu, *Space Tourism*, SpringerBriefs in Applied Sciences and Technology,
https://doi.org/10.1007/978-981-92-0883-8_13

However, for manned spaceflight, increasing passenger numbers introduces compounded risk factors requiring careful trade-off analysis.

Currently, three leading technology solutions are offered by Virgin Galactic, Blue Origin, and SpaceX, respectively. Virgin Galactic's air-launch system and Blue Origin's vertical launch system are fully reusable, but their thrust cannot reach the orbital velocity and can only be used for suborbital tourism.

SpaceX's Starship features a 100-passenger capacity and a 30+ engine cluster with full reusability. Additionally, to reduce the multistage rocket to a two-stage system and enhance the capacity of the crewed Starship for interplanetary travel, such as to Mars, several dedicated tanker flights, also reusable, can be launched separately to refuel the Starship in orbit. Starship's primary risk lies in its high passenger capacity, which could amplify potential disaster consequences.

For the development of manned spaceship to and from Earth orbit, six-person capsules developed by the government with mature technologies may achieve further cost reductions through mass production and reuse. Low-cost reentry capsules led by commercial space companies are also under development. This can be considered an alternative option for manned space tourism.

Other low-cost technologies for entering space are also being tested, such as methane propulsion. There are some advantages for methane, such as low-cost, easy storage and refueling, and it does not require cleaning after use. Despite its lower specific impulse compared to liquid hydrogen and oxygen engines, its apparent advantages are driving many commercial space companies to conduct research and trials on methane propulsion. It may soon enter practical application in the near future. It is estimated that within a decade, per-kilogram crewed launch costs will likely fall below $1000 and continue declining with market expansion.

13.2 ISRU on the Moon and Small Bodies

For the Earth-Moon transportation systems, particularly those supporting lunar tourism, the fuel supply should not rely solely on the Earth. As spaceships routinely travel from Earth orbit to lunar orbit and lunar surface, developing lunar fuel extraction technology becomes an imperative. The immediate priority is achieving breakthroughs in robotic fuel production from lunar regolith.

While extracting fuel from water ice, as discussed in Chap. 7, is technically simpler, the limited distribution of lunar water ice makes regolith-based fuel production more promising. Therefore, the lunar surface infrastructure can combine fuel extraction, tourist activities, and space hotel construction. Thus, fuel extraction facilities should integrate with tourism infrastructure, including take-off and landing sites and communication and tracking networks. Furthermore, refueling stations for Mars and deep space missions can be established in lunar orbit or at the Earth-Moon L1 Lagrange point.

Thus, developing lunar fuel extraction capabilities is equally critical for interplanetary tourism. While fuel transfer and docking technologies are already mature, urgent progress is needed in extraction technologies and the production process.

13.3 Government Policies and Legal Frameworks

On July 23, 2020, the US government released its strategic paper "*A New Era for Deep Space Exploration and Development*", explicitly opening Low-Earth Orbit (LEO) space station operations to commercial entities. This followed years of supporting low-cost commercial launch vehicle development to launch government missions. By commercializing LEO stations, the USA has created a complete market ecosystem for orbital tourism, a significant shift from the government-dominated tourism at the International Space Station (ISS).

As we have seen, due to the fact that the construction and maintenance of the ISS are entirely within the realm of government space programs, the costs remain high. Once the LEO space station opens up to commercial space companies, it will undoubtedly motivate private enterprises to enter and build cost-effective LEO space hotels.

In contrast, Europe and China have moved more cautiously in this aspect. In 2016, China began gradually opening its space sector to commercial companies to allow them to undertake government missions. On May 30, 2019, China released a notice on promoting the orderly development of commercial launch vehicles, permitting commercial rocket development. But for space tourism, China currently lacks specific policies to guide the development of this emerging industry.

As an intergovernmental space agency, ESA has actively supported commercial space industrialization. In 2020, it co-established the European Space Resources Innovation Center (ESRIC) with the Luxembourg government.

Russia's approach focuses on utilizing existing assets like Soyuz spacecraft and ISS to create revenue, but it lacks substantive efforts to reduce costs or expand market accessibility.

The legal status of space resources, including resource utilization and commercialization, remains contentious at the United Nations Office for Outer Space Affairs (UNOOSA). *The Outer Space Treaty, formally known as the Treaty on Principles Governing the Activities of States in the Exploration and Use of Outer Space, Including the Moon and Other Celestial Bodies*, adopted in 1967, declares outer space is a common interest of mankind, implicitly prohibiting national or private appropriation. However, this treaty only stipulates principles and does not provide specific actions. Critical questions remain, such as can companies profit from lunar fuel extraction and orbital refueling services?

The USA bypassed this ambiguity through its 2015 *Commercial Space Launch Competitiveness Act*, explicitly legalizing space resource commercialization. It supports "commercial exploration for and commercial recovery of space resources by

United States citizens".[1] This bold move catalyzed private investment into the commercial space sector.

Whether they are United Nations treaties or the laws of national governments, the highest principle and main purpose should be to protect the space environment and resources and to accelerate the pace of mankind's entry into space.

Despite the fact that the United Nations has not reached a consensus to renew the treaty, it is understandable for governments of various countries, motivated and guided by this principle, to initiate their own policies and laws proactively to gain a first-mover advantage.

The primary goal of creating policies and laws to safeguard space resources is to ensure fair and equitable development. However, if these policies and laws impede mankind's access to space and hinder economic development, or even become an obstacle to mankind's progress, they must be suspended and eventually discarded.

In addition to the legal issues related to the ownership of space resources, for space tourism, there is also the issue of personnel management, i.e., after entering space, what will be the relationship between these passengers? Who will maintain order and enforce the laws of which country?

For example, in a spaceship of a nationally registered company, which law should be enforced in case of incidents or disputes? In a space hotel or a lunar hotel, if a criminal case occurs, what law should be enforced and who should enforce the law? Therefore, prior to massive space tourism, preemptive international legal frameworks are essential to avoid protracted jurisdictional conflicts among countries [1].

Commercial space companies keenly hope to receive government policy support and legal protection in various aspects. This not only motivates investors, protects their interests, and further enhances their confidence, but also orchestrates the development of commercial space companies engaging in space tourism, regulates their businesses, and protects the interests of future tourists (consumers).

Presently, only the USA has established comprehensive space tourism regulations, while other major spacefaring nations lag behind in research on related policies and laws. This will inevitably slow down the progress of space tourism development and even slow down the process of mankind's venture into the solar system, transforming ourselves from a two-dimensional species living on the surface of the Earth to a three-dimensional species extending our activities into space.

13.4 Public Outreach

Public outreach, or science popularization and education, is as important as government policies and laws. In the 1960s, supported by a transparent and open science communication policy in the USA, the Apollo program inspired multiple

[1] U.S. COMMERCIAL SPACE LAUNCH COMPETITIVENESS ACT [2015-11-25] https://www.congress.gov/114/plaws/publ90/PLAW-114publ90.pdf

generations, cultivating America's distinctive pioneering spirit in space exploration. This transformative impact of the Apollo program is undeniable.

When the crew of the Apollo 8 mission captured the iconic *Earthrise* photograph during humanity's first lunar orbiting mission, it captivated the Western world and provided the second enlightenment from space (see Chap. 3). It is a perfect example to demonstrate the great influence of public outreach.

While market forces and technological progress are essential for humanity's adventures beyond the Earth, they cannot succeed without parallel investments in public outreach and education. It is precisely the curious, adventurous individuals, particularly the social elite, who will be the first wave of space tourists, representing a group with brand-new cosmic visions.

Among all science communication mediums, science fiction uniquely ignites imagination and curiosity. *2001: A Space Odyssey*, a landmark science fiction novel by American writer Arthur Clarke, was produced almost simultaneously with the Apollo program in the 1960s. The movie *2001: A Space Odyssey* came out even before Apollo 11's moon landing.

Such rigorous "hard-core" science fiction, blending scientific accuracy with visionary technologies, plays an irreplaceable role in inspiring new generations of space pioneers. Before the arrival of large-scale space tourism, it is also one of the imminent tasks to vigorously encourage the creation and publication of popular science works and hard-core science fiction works on the theme of mankind's access to space [2].

References

1. L.J. Smith, I. Baumann, S.-G. Wintermuth (eds.), *Routledge Handbook of Commercial Space Law* (Routledge, 2025)
2. J. Wu, *L'hôtel de la mer des pluies* [The Lunar Hotel] (EDP Sciences, 2021)

The manufacturer's authorised representative in the EU is Springer Nature Customer Service Centre GmbH, Europaplatz 3, 69115 Heidelberg, Germany. If you have any concerns regarding our products, please contact ProductSafety@springernature.com

Printed and bound by CPI Group (UK) Ltd, Croydon, CR0 4YY

15/09/2026

02216997-0001